中国海洋统计年鉴

CHINA MARINE STATISTICAL YEARBOOK 2012

国 家 海 洋 局

Edited by
State Oceanic Administration,
People's Republic of China

海洋出版社

China Ocean Press

图书在版编目（CIP）数据

中国海洋统计年鉴．2012/国家海洋局编著．--北京：海洋出版社，2013.4

ISBN978-7-5027-8521-5

Ⅰ．①中… Ⅱ．①国… Ⅲ．①海洋—统计资料—中国—2012—年鉴Ⅳ．①P7-66

中国版本图书馆 CIP 数据核字（2013）第 052433 号

责任编辑：王　溪

责任印刷：刘志恒

海洋出版社 出版发行

http: //www.oceanpress.com.cn

北京市海淀区大慧寺路 8 号　　邮编：100081

国家海洋信息中心印刷厂印刷　　新华书店发行所经销

2013 年 4 月第 1 版　　2013 年 4 月第 1 次印刷

开本：787mm×1092mm　1/16　印张：19.75

字数：495 千字

定价：168.00 元

发行部：(010)62147016　　邮购部：(010)68038093　　总编室：(010)62114335

《中国海洋统计年鉴》编委会

周　忠　　中国船舶工业集团公司
刘　悦　　中国船舶重工集团公司
李　军　　中国石油天然气集团公司
刘　岩　　中国石油化工集团公司
王中安　　中国海洋石油总公司
胡红江　　中国盐业总公司
王华俊　　中国有色金属工业协会
杨宝瑞　　辽宁省海洋与渔业厅
王保民　　河北省国土资源厅
孙连友　　天津市海洋局
王守信　　山东省海洋与渔业厅
唐庆宁　　江苏省海洋与渔业局
沈依云　　上海市海洋局
陈宗尧　　浙江省海洋与渔业局
张福寿　　福建省海洋与渔业厅
洪伟东　　广东省海洋与渔业局
杨小光　　广西壮族自治区海洋局
潘建纲　　海南省海洋与渔业厅
栾玉瑄　　大连市海洋与渔业局
黄聿颂　　青岛市海洋与渔业局
吴建义　　宁波市海洋与渔业局
王春生　　厦门市海洋与渔业局
梁俊乾　　深圳市海洋局

《中国海洋统计年鉴》编辑部

徐　婧　中国船舶重工集团公司
孙效娴　中国石油天然气集团公司
孙　艳　中国石油化工集团公司
郝静辉　中国海洋石油总公司
杜丽华　中国盐业总公司
孙秀敏　中国有色金属工业协会
魏　南　辽宁省海洋与渔业厅
高　娜　河北省国土资源厅
杨金璐　天津市海洋局
苏庆猛　山东省海洋与渔业厅
裴　沛　江苏省海洋与渔业局
张　呈　上海市海洋局
应　晓　浙江省海洋与渔业局
潘志敏　福建省海洋与渔业厅
利红兵　广东省海洋与渔业局
何　精　广西壮族自治区海洋局
潘　骏　海南省海洋与渔业厅
崔　楠　大连市海洋与渔业局
王子江　青岛市海洋与渔业局
韩　静　宁波市海洋与渔业局
池信才　厦门市海洋与渔业局
罗小霞　深圳市经济贸易和信息化委员会

执行编辑：李长如

英文校订：林宝法

Editorial Board of China Marine Statistical Yearbook

Cao Ning	Chinese Academy of Sciences
Xu Tieju	China Earthquake Administration
Wang Bangzhong	China Meteorological Administration
Zhou Zhong	China State Shipbuilding Corporation
Liu Yue	China Shipbuilding Industry Corporation
Li Jun	China National Petroleum Corporation
Liu Yan	China Petrochemical Group Corporation
Wang Zhong'an	China National Offshore Oil Corporation
Hu Hongjiang	China National Salt Industry Corporation
Wang Huajun	China Nonferrous Metals Industry Association
Yang Baorui	Department of Ocean and Fisheries of Liaoning Province
Wang Baomin	Hebei Province Department of Land and Resources
Sun Lianyou	Tianjin Ocean Administration
Wang Shouxin	Department of Oceanic and Fishery of Shandong Province
Tang Qingning	Jiangsu Provincial Ocean and Fisheries Bureau
Shen Yiyun	Shanghai Ocean Administration
Chen Zongyao	Zhejiang Provincial Ocean and Fisheries Bureau
Zhang Fushou	Department of Oceanic and Fishery of Fujian Province
Hong Weidong	Guangdong Provincial Oceanic and Fishery Administration
Yang Xiaoguang	Guangxi Ocean Administration
Pan Jiangang	Department of Marine and Fishery of Hainan Province
Luan Yuxuan	Dalian Ocean and Fishery Administration
Huang Yusong	Qingdao Ocean and Fishery Administration
Wu Jianyi	Ningbo Ocean and Fishery Bureau
Wang Chunsheng	Oceans and Fisheries Bureau of Xiamen
Liang Junqian	Shenzhen Ocean Administration

Editorial Department of China Marine Statistical Yearbook

Wu Wanke	China Meteorological Administration
Qiu Yinlang	China State Shipbuilding Corporation
Xu Jing	China Shipbuilding Industry Corporation
Sun Xiaoxian	China National Petroleum Corporation
Sun Yan	China Petrochemical Group Corporation
Hao Jinghui	China National Offshore Oil Corporation
Du Lihua	China National Salt Industry Corporation
Sun Xiumin	China Nonferrous Metals Industry Association
Wei Nan	Department of Ocean and Fisheries of Liaoning Province
Gao Na	Hebei Province Department of Land and Resources
Yang Jinlu	Tianjin Ocean Administration
Shu Qingmeng	Department of Oceanic and Fishery of Shandong Province
Pei Pei	Jiangsu Provincial Ocean and Fisheries Bureau
Zhang Cheng	Shanghai Ocean Administration
Ying Xiao	Zhejiang Provincial Ocean and Fisheries Bureau
Pan Zhimin	Department of Oceanic and Fishery of Fujian Province
Li Hongbing	Guangdong Provincial Oceanic and Fishery Administration
He Jing	Guangxi Ocean Administration
Pan Jun	Department of Marine and Fishery of Hainan Province
Cui Nan	Dalian Ocean and Fishery Administration
Wang Zijiang	Qingdao Ocean and Fishery Administration
Han Jing	Ningbo Ocean and Fishery Bureau
Chi Xincai	Oceans and Fisheries Bureau of Xiamen
Luo Xiaoxia	Shenzhen Municipal Economic, Trade and Informanization Commission

Executive Editor: Li Changru

English Proof-reader: Lin Baofa

编 者 说 明

一、《中国海洋统计年鉴》2012 年版是一部全面反映 2011 年中华人民共和国海洋经济发展和海洋管理服务情况的资料性年鉴，全书为中英文对照。

二、本年鉴的统计资料范围为人们在海洋和沿海地区开发、管理、利用海洋资源和空间，发展海洋经济的生产和活动以及沿海地区的社会经济概况。地域范围为沿海地区、沿海城市和沿海地带，其排列顺序按《沿海行政区域分类与代码》(HY/T 094-2006) 的顺序排列。

三、本年鉴内容包括综合资料、海洋经济核算、主要海洋产业活动、主要海洋产业生产能力、涉海就业、海洋科学技术、海洋教育、海洋环境保护、海洋行政管理及公益服务、全国及沿海社会经济、部分世界海洋经济统计资料等十一部分。

四、本年鉴根据《海洋统计报表制度》（国统制[2012]50 号）和《海洋生产总值核算制度》（国统制[2011]7 号），资料主要来源于沿海省、自治区、直辖市统计局、海洋厅（局）以及 20 个有关涉海部、局、总公司。

五、本年鉴中除特殊说明外，所有价值量指标均为当年价，除注明年份外，其他均为 2011 年数据，年鉴中每部分后附有主要海洋统计指标解释，对主要海洋统计指标的含义、统计范围和统计方法做了简要说明。统计数据中的其他说明置于表的下面。有续表的资料，如有注释均置于最后一张表的下面。

六、本年鉴表格中符号使用说明：“…”表示数据不足本表最小计算单位数；“空格”表示该项统计指标数据不详或无该项数据；“#”表示其中的主要项；其他符号如“*”或“①”等表示本表后面有注释。

七、本年鉴中由于数字精确度的原因，四舍五入后部分分项值之和与合计值有微小差异。

八、本年鉴资料国内部分未包括香港特别行政区、澳门特别行政区和台湾省数据。

九、《中国海洋统计年鉴》在编撰过程中，得到了各有关单位的大力支持，我们在此表示衷心的感谢。本年鉴中如有疏漏和不妥之处，敬请读者批评指正。

《中国海洋统计年鉴》编辑部

Introduction

I. *China Marine Statistical Yearbook (2012)* is a data almanac which reflects in an all-round way the development of marine economy, marine management and service in the People's Republic of China in 2011, and it is a Chinese-English bilingual edition.

II. The Yearbook's statistics cover the production and activities in the marine and coastal areas in relation to the development, management and utilization of marine resources and space, and the development of marine socioeconomy. The regions covered are the coastal regions, coastal cities and coastal zones with coastlines, which are arranged in order according to the *Coastal Administrative Areas Classification and Codes* (HY/T 094—2006).

III. The data in the yearbook consist of 11 sections, namely, integrated data, marine economic accounting, major marine industrial activities, production capacity of major marine industries, ocean-related employment, marine science and technology, marine education, marine environmental protection, marine administration and public-good service, national and coastal socioeconomy, part of the world's marine economic statistics data.

IV. The Yearbook is based on the *Marine Statistics Report System* (Guotongzhi [2012] No. 50) and the *Ocean Gross Product Accounting System* (Guotongzhi [2011] No. 7), its data mainly come from the statistical bureaus and the oceanic administrations of the coastal provinces, autonomous regions, and municipalities directly under the Central Government as well as the 20 ocean-related ministries, bureaus and general corporations concerned.

V. Unless otherwise specified in the Yearbook, all the value indicators are given at the current price. Each section is attached by explanatory notes to the major marine statistical indicators, giving a brief explanation for the meaning, statistical range and statistical methods of the major marine statistical indicators. Other notes to the statistical data are listed below the tables. For the data with continued tables, annotations, if any, are put below the last table.

VI. The usage of symbols in the tables: "…" indicates the statistics smaller than the minimum calculation unit in the table; "Blank" indicates that the data of the statistical index is unknown for the time being or that there shouldn't be any data; "#" indicates the major items of the table; Other symbols, such as "*" or "①", indicate "see footnotes below".

VII. For reasons of digital accuracy, there is small difference between the sum of values of some subterms after having been rounded off and the total values.

VIII. The domestic part of the Yearbook does not include that of Hong Kong Special Administrative Region, Macau Special Administrative Region and Taiwan Province.

IX. In the course of editing *China Marine Statistical Yearbook*, we enjoyed energetic support from the various departments concerned and we hereby extend our heartfelt thanks to them. Criticisms and comments are welcome from readers on any of the oversights and inappropriateness as the time for editing is too short.

Editorial Department of
China Marine Statistical Yearbook

目　次
CONTENTS

3 主要海洋产业活动

Major Marine Industrial Activities

4 主要海洋产业生产能力
Production Capacity of Major Marine Industries

7 海洋教育

Marine Education

8 海洋环境保护

Marine Environmental Protection

2011年我国海洋经济发展综述

2011年，沿海各地区深入贯彻实践科学发展观，认真落实党中央、国务院发展海洋经济的战略部署，积极推进全国海洋经济发展试点工作，加快推进经济发展方式转变和结构调整，海洋经济继续保持平稳较快的发展势头，实现了“十二五”时期的良好开局。

一、全国海洋经济发展概况

2011年，全国海洋生产总值45 496.0亿元，比2010年增长9.89%（除特别注明外，增长率均按可比价计算），海洋生产总值占国内生产总值的9.62%，占沿海地区生产总值的15.7%。全国涉海就业人员3 421.7万人，比2010年增加70.9万人。

二、主要海洋产业发展情况

2011年主要海洋产业实现增加值18 865.2亿元，比2010年增长9.1%，占海洋生产总值的41.5%，滨海旅游业和海洋交通运输业仍占主导地位。

海洋第一产业 2011年，沿海地区海洋渔业生产稳定，海洋水产品产量稳步增长，远洋渔业综合实力不断增强。海水养殖面积210.6万公顷，海洋渔业全年实现增加值3 202.9亿元，按可比价计算，比2010年增长1.1%；海水产品产量达2 908.0万吨，比2010年增长4.0%，其中，海水养殖产量1 551.3万吨，比2010年增长4.7%；海洋捕捞产量1 241.9万吨，比2010年增长3.2%；远洋捕捞产量114.8万吨，比2010年增长29.3%。

海洋第二产业 2011年，海洋第二产业继续保持良好的发展态势。随着沿海多个风电场项目相继竣工投产，海洋电力业继续保持快速增长态势，全年实现增加值59.2亿元，比2010年增长52.8%。在国家相关政策的激励下，海洋生物医药业快速增长，全年实现增加值150.8亿元，比2010年增长21.3%。随着国家扶持政策的陆续出台以及多项技术的重大突破，我国海水利用产业规模不断扩大，全年实现增加值10.4亿元，比2010年增长11.9%。海洋船舶工业继续保持平稳发展态势，全年实现增加值1 352.0亿元，比2010年增长10.8%，造船完工量达3 412艘，比2010年增长42.9%。海洋盐业生产平稳发展，全年海盐产量达3 322.4万吨，比2010年增长1.1%，实现增加值76.8亿元，比2010年增长7.4%。海洋工程建筑业实现增加值1 086.8亿元，比2010年增长13.9%。受溢油等突发事件影响，海洋原油产量有所下降，但随着油气价格的上涨，海洋油气业依然保持了稳定发展，全年实现增加值1 719.7亿元，比2010年增长6.0%，全年海洋原油产量4 452万吨，比2010年减少5.5%，海洋天然气产量1 214 519万立方米，比2010年增长9.5%。海洋化工业全年实现增加值695.9亿元，比2010年增长3.2%。海洋矿业开采品种不断丰富，产量继续保持平稳增长，全年实现增加值53.3亿元，比2010年增长2.6%。

海洋第三产业　2011年，全国沿海港口生产态势总体良好，但航运业受国际需求放缓及航运价格下跌等因素影响仍处于低迷状态。全年海洋交通运输业实现增加值4 217.5亿元，比2010年增长10.8%，沿海港口货物吞吐量达636 024万吨，比2010年增长12.7%，国际标准集装箱吞吐量14 632万标准箱，比2010年增长11.3%。沿海地区以当地滨海旅游资源为依托，规划滨海旅游产业的发展，支持滨海旅游项目的上马，注重滨海旅游设施的完善，加强滨海旅游产品的创新，积极创造良好发展环境，滨海旅游业全年实现增加值6 239.9亿元，比2010年增长12.1%。

三、区域海洋经济发展情况

2011年，环渤海、长三角、珠三角三大经济区海洋经济继续保持平稳增长的态势，海洋经济总量再创新高，但随着国家宏观调控政策的实施，沿海省份均将工作重点转移到“转方式、调结构”上来，虽然增速均表现出不同程度的回落。环渤海经济区、长江三角洲经济区和珠江三角洲经济区海洋生产总值分别为16 345.2亿元、14 408.4亿元和9 191.1亿元，占全国海洋生产总值的比重分别为35.9%、31.7%、20.2%。

环渤海经济区海洋经济增速放缓，海洋生产总值比2010年增长17.9%（现价），占地区生产总值比重15.8%。海洋产业增加值为9 478.5亿元，海洋相关产业增加值为6 866.7亿元。与上年相比各项主要海洋产业均实现增长，海洋渔业、海洋油气业、海洋交通运输业、滨海旅游业四大海洋支柱产业增加值合计达到6 365.4亿元，占该地区主要海洋产业增加值的83.8%。海洋生物医药业、海洋电力业两大海洋新兴产业增幅明显，分别与上年增长149.4%和71.6%（现价）。

长江三角洲经济区增速继续减缓，海洋生产总值比2010年增长13.8%（现价），海洋生产总值占地区生产总值比重14.3%。长江三角洲海洋产业增加值8 210.5亿元，海洋相关产业增加值6 197.9亿元。按照产值贡献，滨海旅游业、海洋交通运输业、海洋船舶工业和海洋渔业四个产业位居前列，其增加值之和占该地区主要海洋产业增加值的92.1%，其中滨海旅游业占该地区主要海洋产业增加值的41.1%，产值贡献位居第一。按照增长速度，海洋油气业继续保持高速的增长速度，增长速度达91.0%（现价），海洋电力业、海洋工程建筑业等海洋新兴产业紧随其后，增长速度分别达到32.1%和23.4%（现价）。

2011年，珠江三角洲经济区海洋经济总量稳步增长，海洋生产总值达9 191.1亿元，比2010年增长11.4%（现价），海洋生产总值占地区生产总值比重达17.3%。海洋产业增加值为5 612.2亿元，海洋相关产业增加值为3 578.9亿元。滨海旅游业、海洋交通运输业、海洋油气业、海洋化工业和海洋渔业依然为珠江三角洲经济区海洋经济发展的支柱产业，其增加值之和占该地区主要海洋产业增加值92.4%。海洋电力业增长迅速，增长速度达到50.4%（现价）。

四、海洋科研教育

2011年，海洋科研教育事业继续保持稳步发展，海洋科研机构共179个，从业人员37 445人，比2010年增长5.8%；海洋科研机构承担海洋科技课题14 253项，比2010年增长5.8%；发表海洋科技论文15 547篇，出版海洋科技著作278种，分别比2010年增长8.8%和9.4%；专利授权数2 034件，其中发明专利1 355件，分别比2010年增长37.2%和37.1%。开设海洋专业的高等学校共346个，比去年增加5.8%，专任教师254 042人，比2010年增加14 387人；高等教育和中等职业教育海洋专业毕业生人数共97 469人，比2010年增加6.1%，招生人数和在校生人数分别为100 316人和308 798人，硕士以上研究生专业点数比2010年增加4个。

五、海洋环境保护

2011年，我国海洋环境状况总体维持在较好水平。符合第一类海水水质标准的海域面积约占我国管辖海域面积的95%，海洋沉积物质量良好，浮游生物和底栖生物的生物多样性及群落结构基本稳定。海水、海洋沉积物、海洋生物的放射性水平和海洋大气γ辐射空气吸收剂量率均处于本底范围内。海洋保护区环境状况总体良好，主要保护对象或保护目标基本保持稳定。海水浴场、滨海旅游度假区等旅游休闲娱乐区水质总体良好。海水增养殖区环境质量基本满足养殖活动要求。海洋倾倒区环境可满足继续使用的要求。但是，我国近岸海域环境问题仍然突出，主要表现在陆源排污压力巨大，近岸海域污染严重，赤潮灾害多发，局部区域海水入侵、土壤盐渍化、海岸侵蚀等灾害严重，海洋溢油等突发性事件的环境风险加剧等。2011年我国共发生114次风暴潮、海浪和赤潮过程，其中44次造成灾害。各类海洋灾害（含海冰、涌潮等）造成直接经济损失62.07亿元，死亡（含失踪）76人。

六、海洋行政管理

2011年，海洋综合管控能力稳步提高，行政管理工作有序开展。全年颁发海域使用权证书3 874本，比2010年增长56.1%；确权海域面积18.59万公顷，比2010年略有减少；全年共签发疏浚物海洋倾倒许可证95份，实施各项海洋执法检查共145 598次，发现违法行为2 113起，比2010年增长15.1%；提供海洋数值预报服务共12 519次，开展海洋调查项目175个，获得数据共2 072.29万个，各项海洋观测获得数据量共19 457.76万个，全年接收存档卫星遥感数据量共计33 828.24 GB；本年接收纸质档案8 337卷（册），电子档案788 GB；共有25项行业标准通过立项审查，发布国家标准1项，行业标准11项。

Summary of China's Marine Economic Development in 2011

In 2011, all coastal regions intensively carry out and practice the concept of scientific development, conscientiously put into effect the strategic plan of the Central Party Committee and the State Council for the development of marine economy, actively give impetus to the pilot work on national marine economic development, speed up the advancement of the transformation of the pattern of economic development and structural readjustment, so that the marine economy continues to deep a momentum of stable and rapid growth and realize a good start of the 12th Five-year Plan period.

I. Survey of the National Marine Economic Development

In 2011, the gross ocean product of national marine economy reaches 4 549.6 billion yuan (RMB), 9.89% up from the previous year (Unless otherwise specified, the growth rate is calculated at the comparable price.), its poportious in the GDP and the Gross Regional Product being 9.62% and 15.7% respectively. The number of people employed by the ocean related sectors throughout the country amounts to 34.217 million, 0.709 million more than that in the previous year.

II. Development of Major Marine Industries

Development of major marine industries realize an added ralue of 1 886.52 billion yuan, growing by 9.1% as against that of the previous year, accounting for 41.5% of the national gross ocean product, and coastal tourism and marine communications and transportation still occupy the leading position.

Primary marine industry

In 2011, the marine fishery production in the coastal region keeps stable, the output of marine aquatic products grows steadily and the integrated strength of deep-sea fishing is contantly improved. The area of maricalture. Amounts to 2.106 million ha, and the full-year marine fishery effects an added value of 320.29 billion yuan, which, calculated at the comparable price, increases by 1.1% as compared with that in the previous year; the yield of marine aquatic products reaches 29.08 million tons, 4.0% up from the previous year, among which, the output of mariculture is 15.513 million tons, 4.7% up from the previous year; the production of marine fishing 12.419 million tons, 3.2% up from the precious year and that of deep-sea fishing 1.148 million tons, 29.3% up from the previous year.

Secondary marine industry

In 2011, the secondary marine industry continues to keep a good posture of development. With a number of coastal wind power generation field projects having been completed and put into operation one by one, the marine electric power industry continues to maintain and posture of rapid growth, effecting an added value of 5.92 billion yuan for the whole year, 52.8% up from the previous year. Prompted by the related national policies, the marine biomedicine industry has grown rapidly, accomplishing a full-year added value of 15.08 billion yuan, 21.3% up from the previous year. Along with the promulgation of the national support policies successively as well as major breakthrough in a number of technologies, China's seawater utilization industry is increasingly expanding, realizing a full-year added value of 1.04 billion yuan, increasing by 11.9% as against that in the previous year. The marine shipbuilding industry continues to keep a posture of smooth and steady development, effecting for the whole year an added value of 135.20 billion yuan, 10.8% up from the previous year and the completed quantity of ships built tops by 3 412, 42.9% up from the previous year. The production of sea salt industry is developing smoothly, with the full-year sea salt output reaching 33.224 million tons, 1.1% up from the previous year, effecting an added value of 7.68 billion yuan, 7.4% up from the previous year. The marine engineering architecture industry realizes an added value of 108.68 billion yuan, 13.9% up from the previous year. Affected by unexpected incidents such as oil spill, the yield of marine crude oil has dropped to some extent, but with the rising of the oil and gas price, the offshore oil and gas industry has still maintained a stable evelopment, effecting a full-year added value of 171.97 billion yuan, 6.0% up from the previous year; The output of marine crude oil for the whole year amounts to 44.52 million tons, 5.5% down from the previous year, and that of marine natural gas 12.145 19 billion m^3, 9.5% up from the previous year. The marine chemical industry effects a full-year added value of 69.59 billion yuan, 3.2% up from the previous year. The marine mining industry has exploited increasingly rich varieties of ore and the output continues to keep a smooth and steady growth, effecting a full-year added value of 5.33 billion yuan, 2.6% up from the previous year.

Tertiary marine industry

In 2011, the production situation of national coastal harbours is on the whole fine, but the shipping industry is still in a depressed state as a result of the impact of such factors as relax of the international demand and drop of the shipping price, etc. the marine communications and stransportation industry effects a full-year added value of 421.75 billion yuan, 10.8% up from the previous year, the cargo handling capacity of coastal harbours reaches 6.36 billion tons, 12.7% from the previous year and the handling capacity of international standardized containers 146.32 million standard cases, 11.3% up from the previous year. With the backing of the local resource of coastal tourism, the coastal regions plan for the development of coastal tourist industry, support the launching of coastal tourist projects, pay attention to the perfection of coastal tourist facilities, strengthen the innovation of coastal tourist products and actively create and atmosphere of good development, as a result of which the coastal tourism effects a full-year added value of 623.99 billion yuan, 12.1% up from the previous year.

III. Development of Regional Marine Economy

In 2011, the marine economy in the three major economic zone round the Bohai Sea, in the Changjiang River Delta and the Zhujiang River Delta continues to maintain a posture of smooth and steady development and the toatal economic capacity bits a new high again, but, with the implementation of the national policy of mocrocontrol, all the coastal provinces have shifted their focus of work onto the “Changing the mode and adjusting the structure” despite a drop to varying degrees in the growth rate. The gross ocean products of the Round-the-Bohai Economic Zone, Chingjiang River Delta and Zhujiang River Delta are 1 634.52 billion yuan, 1 440.84 billion yuan and 919.11 billion yuan respectively, accounting for 35.9%, 31.7% and 20.2% of the national gross ocean product separately.

The marine economic growth rate of the Round-the-Bohai Economic Zone has slowed down and the gross ocean product has increased by 17.9% as compared with that last year (at the current price), occupying a proportion of 15.8% in the Gross Regional Product, the added value of marine industries is 947.85 billion yuan and that of the marine related industries 686.67 billion yuan. In comparison with those in the previous year, all the major marine industries have effected a growth and the added value of four major marine industries of marine fishery, marine offshore oil and gas, communications and transportation, and coastal tourism totals 636.54 billion yuan, accounting for 83.8% of the added value of major marine industries in the region. The two major marine emerging industries: marine biomedicine industry and marine electric power industry, have registered a great increase, 149.4% and 71.6% up from that of last year respectively.

The growth rate of the Changjiang River Delta Economic Zone continues to slow down, and the gross ocean product increases by 13.8% (at the current price) as against that last year, accounting for a proportion of 14.3% in the gross regional product. The added value of marine industries in the Changjiang River Delta is 821.05 billion yuan and that of the marine related industries 619.79 billion yuan. In the light of the output value contributions, the industries of coastal tourism, marine communications and transportation, marine shipbuilding and marine fishery rank among the first and the sum total of their added value occupies 92.1% of that of the major maine industries in the region, among which, the added value of coastal tourism accounts for 41.4% of that of major marine industries in the region and its output value contribution ranks first. In light of the growing speed, the offshore oil and gas industry continues to keep a rapid growth rate, which reaches 91.0% (at the current price), following which, are the newly emerging marine industries such as marine electric power and marine engineering architecture with a growth rate of 32.1% and 23.4% respectively (at the current price).

In 2011, the total marine economic capacity of the Zhujiang River Delta economic Zone has steadily grown. The gross ocean product of this zone this year reaches 919.11 billion yuan, 11.4% up from that of last year (at the current price), and the gross ocean product occupies a proportion of 17.3% of the gross regional product. The added value of marine industries is 561.22 billion yuan and that of marine related industries 357.89 billion yuan. Coastal tourism, marine communications and transportations, offshore oil and gas industry, marine chemical industry and marine fishery and still the pillar industries in the marine economic development of the Zhujiang River Delta Economic Zone, the sum total of their added values accounting for 92.4% of the added value of major marine industries in the region. The marine electric power industry grows rapidly at a growth rate of 50.4% (at the current price.)

Ⅳ. Marine Scientific Research and Education

In 2011, the marine scientific research and education continue to develop steadily. There are a total of 179 marine scientific research institutions with 37 445 employees, increasing by 5.8% as compared with last year; the number of marine scientific and technological projects undertaken by there institutions is 14 253, 5.8% up from last year; 15 547 marine scientific and technological papers and 278 kinds of marine scientific and technological works are published, growing by 8.8% and 9.4% respectively as compared with the previous year; the number of patents authorized is 2 034, of which 1 355 is the patents for discovery, increasing by 37.2% and 37.1% respectively as against the previous year. The number of the institutions of higher learning which offer marine specialities reaches

346, 5.8% up from that last year, and that number of full-time teachers is 254 042, 14 387 more than the in the previous year; the number of graduates from the marine speciality of higher leaning and secondary vocational education is 97 469, increasing by 6.1% as compared with that in the previous year, the number of students enrolled and that in school are 100 316 and 308 798 respectively and the number of specialty programmes for graduates above masters is 4 more than that last year.

Ⅴ. Marine Environmental Protection

In 2011, China's marine environmental condition is on the whole maintained at a good level. The area of the sea that is up to the standard of first-class sea water quality accounts for about 95% of the sea area under China's jurisdiction, the quality of marine sediment is fine and the biodiversity of plankton and benthos and their community structure are basically stable. Both the radioactivity level of seawater, marine sediment and marine life and the rate of air absorbed dose of the marine atmospheric radiation are within the background range. The environmental condition of the marine protected area (MPA) is good on the whole and the major objects or targets of protection basically keep stable. The water quality in the tourist leisure and recreational zones such as bathing beach, coastal tourist resorts, etc. is good altogether. The environmental quality in the culture and stock enhancement zone can satisfy the demand for cultivating activities and that in the ocean dumpling zone for further use. But, the environmental problems in the nearshore sea area of China are still prominent and mainly manifested in the huge pressure from terrigenous sewage discharge, serious pollution of the nearshore sea area, frequent occurrence of read tide disasters, serious diseases such as sea water intrusion in local regions, soil stalinization, and the intensification of environmental risks of such sudden incident as maritime oil spill etc. In 2011, a total of 114 storm surge, sea ware and red tide processes, of which 44 have resulted in disaster. All types of marine disasters (including sea ice, swell tide etc.) have caused a direct economic loss of 6.207 billion yuan (RMB) and the death (including missing) of 76 persons.

Ⅵ. Marine Administration

In 2011, the marine integrated capacity of management and control has been improved steadily and the administrative work has been carried out orderly. A total of 3 874 certificates for the sea area use right are issued for the whole year, increasing by 56.1% as against that in the previous years; the sea area with the ownership of patent rights reaches 185 900 hm^2, slightly less than that last year; a total of 95 permits for oceanic dumping of dredged material are signed and issued throughout the year, and marine inspections for law enforcement are carried out on 145 598 occasions in which 2 113 cases of unlawful practice are discovered, increasing by 15.1% as compared with last year; marine numerical forecast service is provided for 12 519 times, 175 marine survey projects are carried out, acquiring 20.722 9 million data and the data quantity for various marine observations totals 194.577 6 million and the quantity of satellite remote-sensing data received and placed on file for the whole year totals 33 828.24 GB; The paper archives received this year total 8 337 volumes (copies); the electronic archives 788 GB. A total of 25 items of professional standard have been arthorized and examined one item of national standard and 11 items of professional standard have been issued.

图1　全国海洋生产总值及三次产业构成
China's Gross Ocean Product and Three Industries Composition

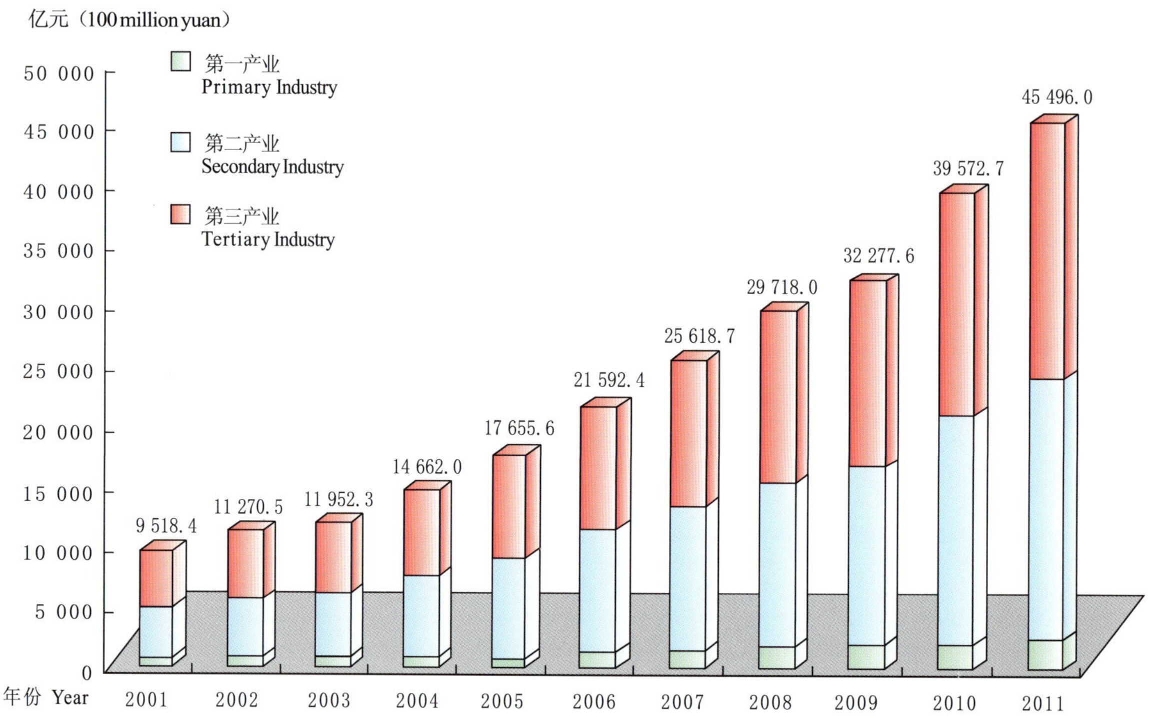

图2　2011年全国主要海洋产业增加值构成
Composition of Added Values of the Major Marine Industries in China in 2011

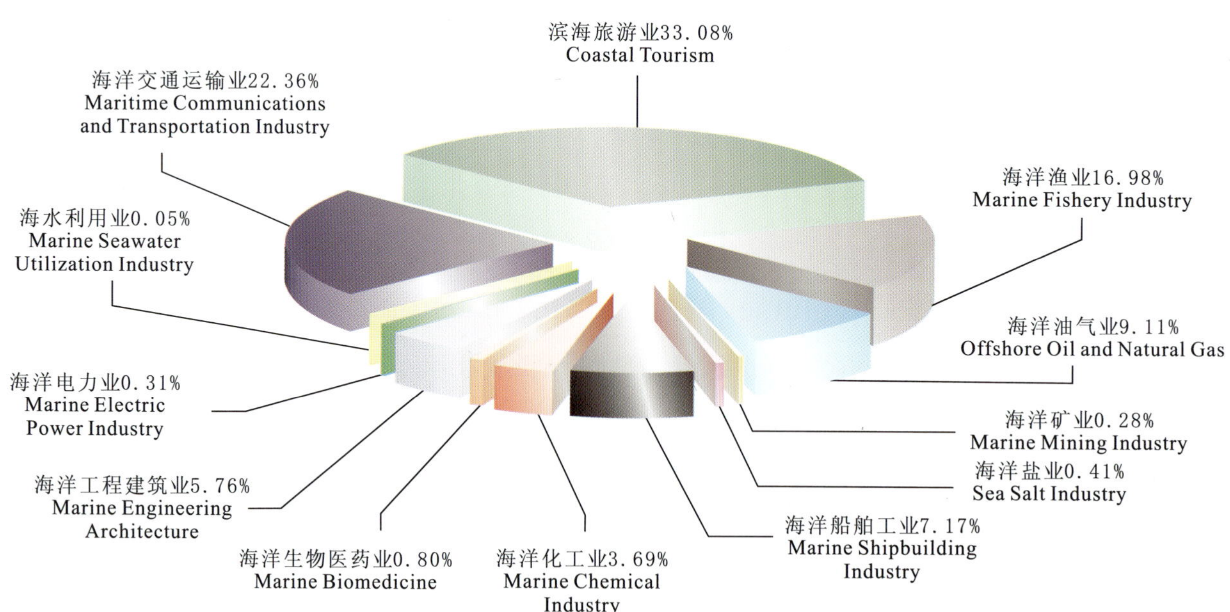

图3 2011年沿海地区海洋生产总值
Gross Ocean Product by Coastal Regions in 2011

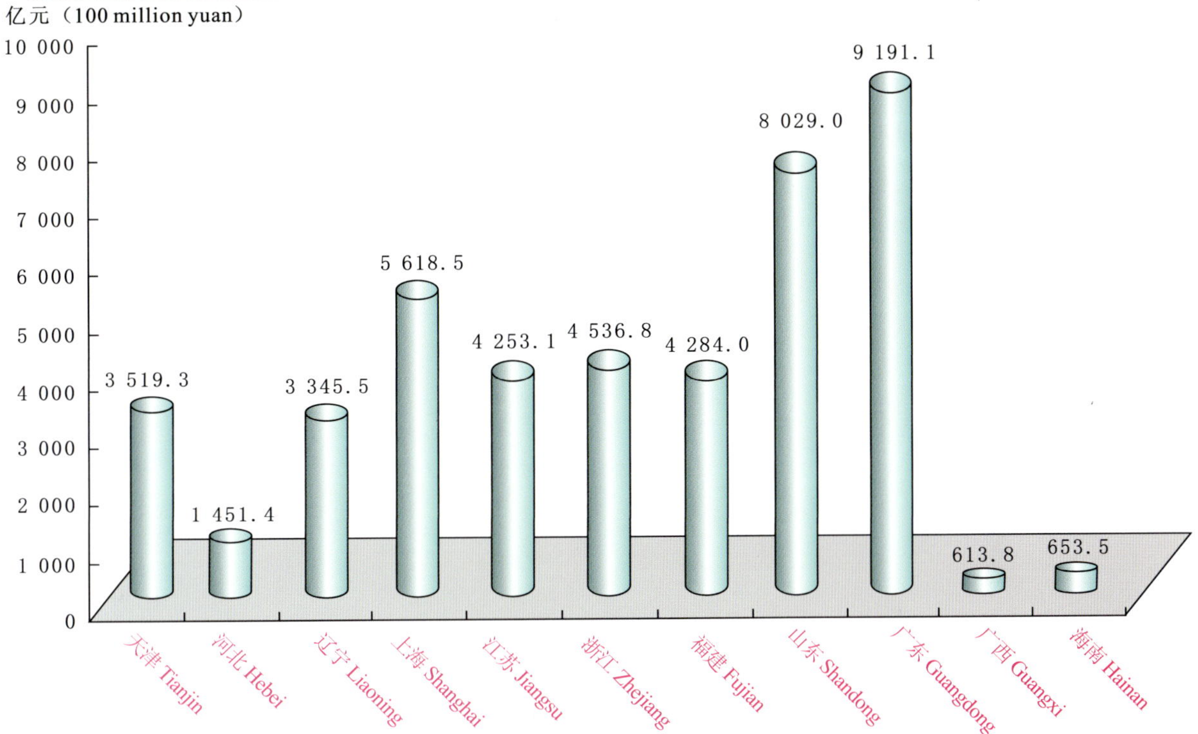

图4 全国海洋捕捞养殖产量
National Marine Catches and Mariculture Production

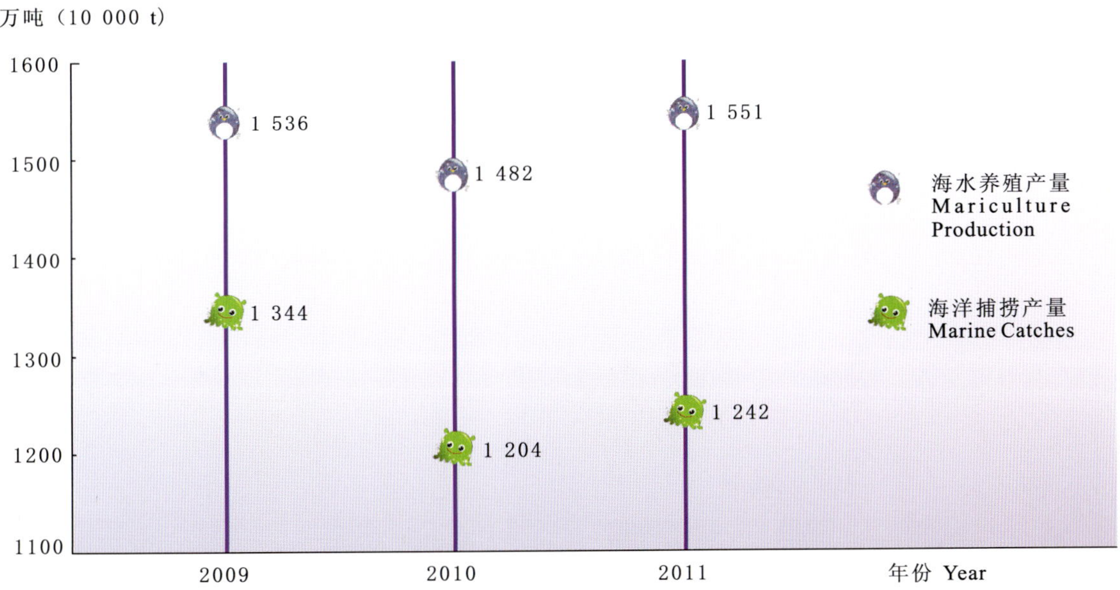

图5 2011年沿海地区海洋捕捞养殖产量

Marine Catches and Mariculture Production by Coastal Regions in 2011

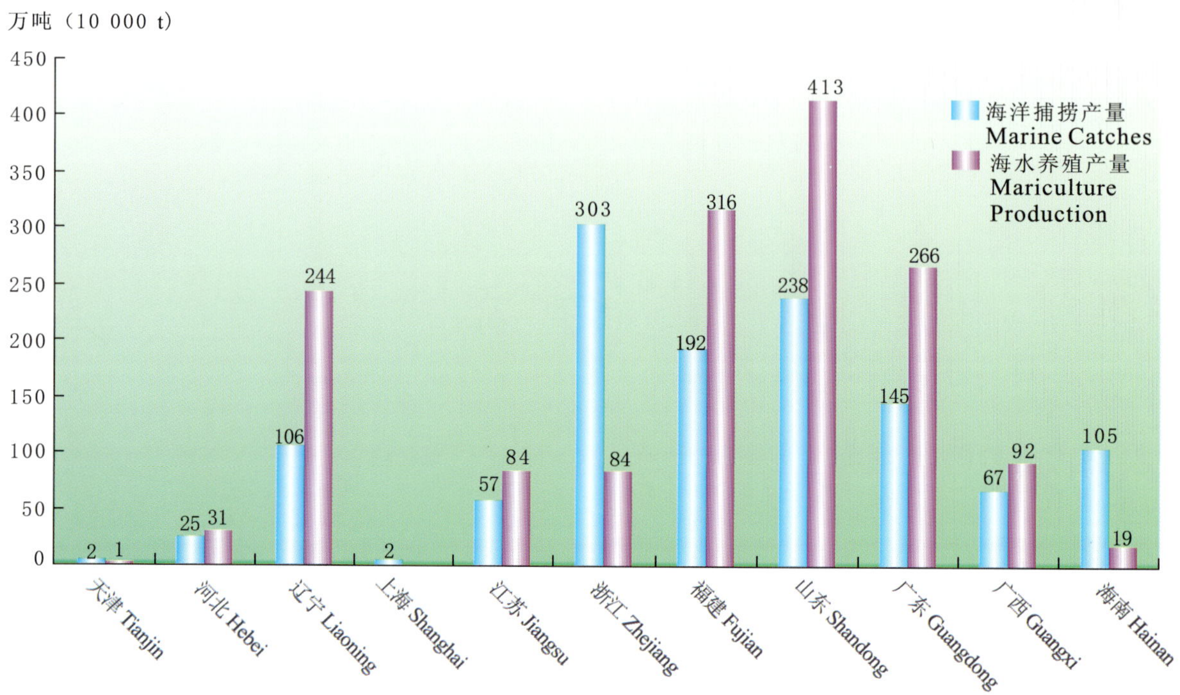

图6 全国海洋石油和天然气产量

National Output of Offshore Oil and Natural Gas

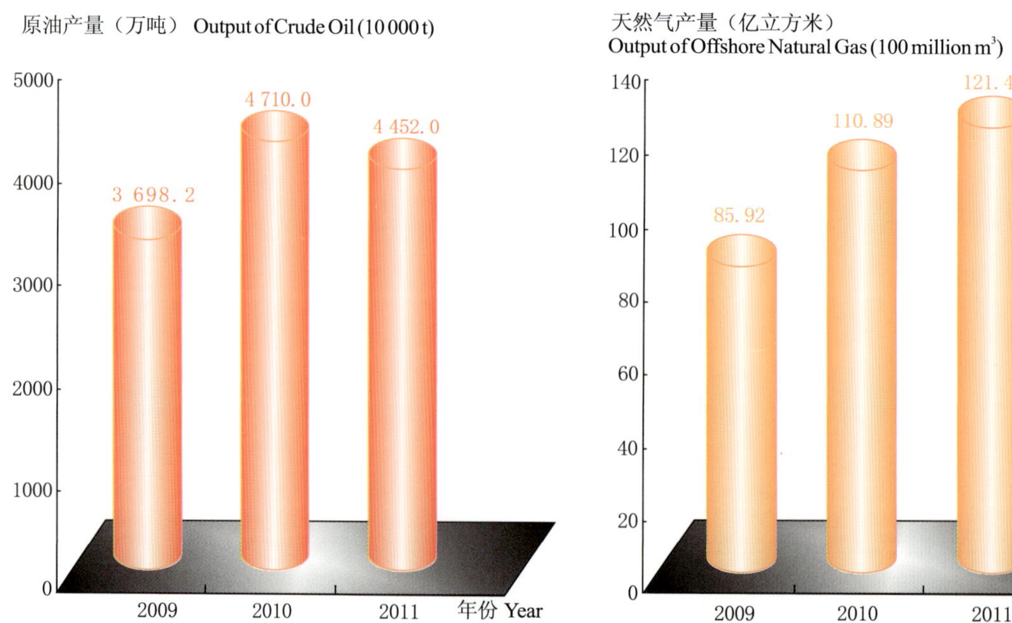

图7 海洋原油产量占全国原油产量比重

Proportion of Offshore Crude Oil Production in the National Total

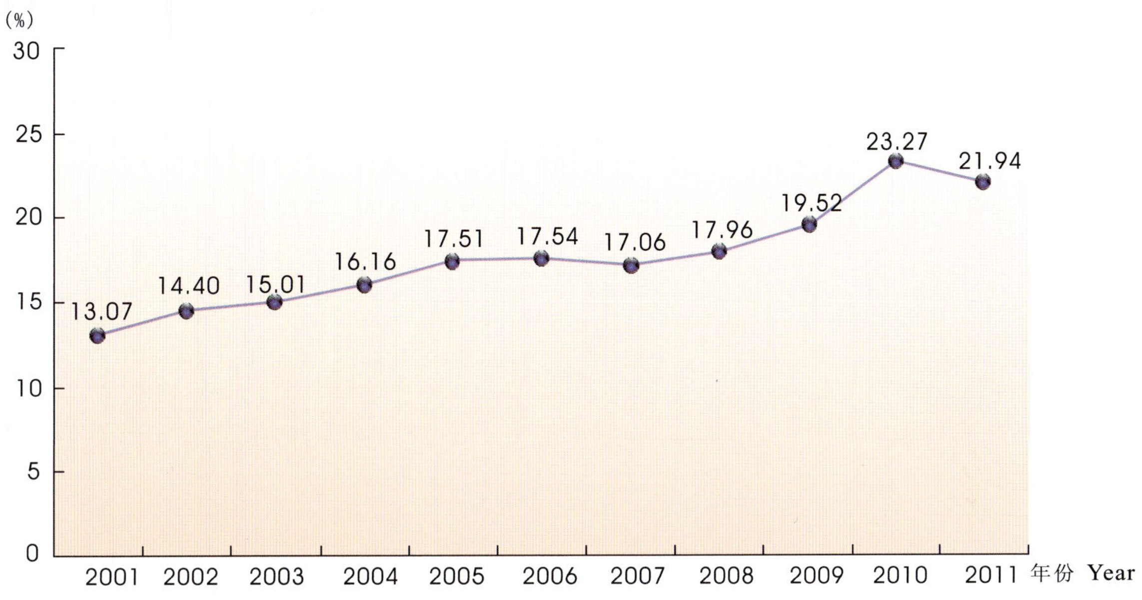

图8 2011年海洋矿业产量

Production of Marine Mining Industry in 2011

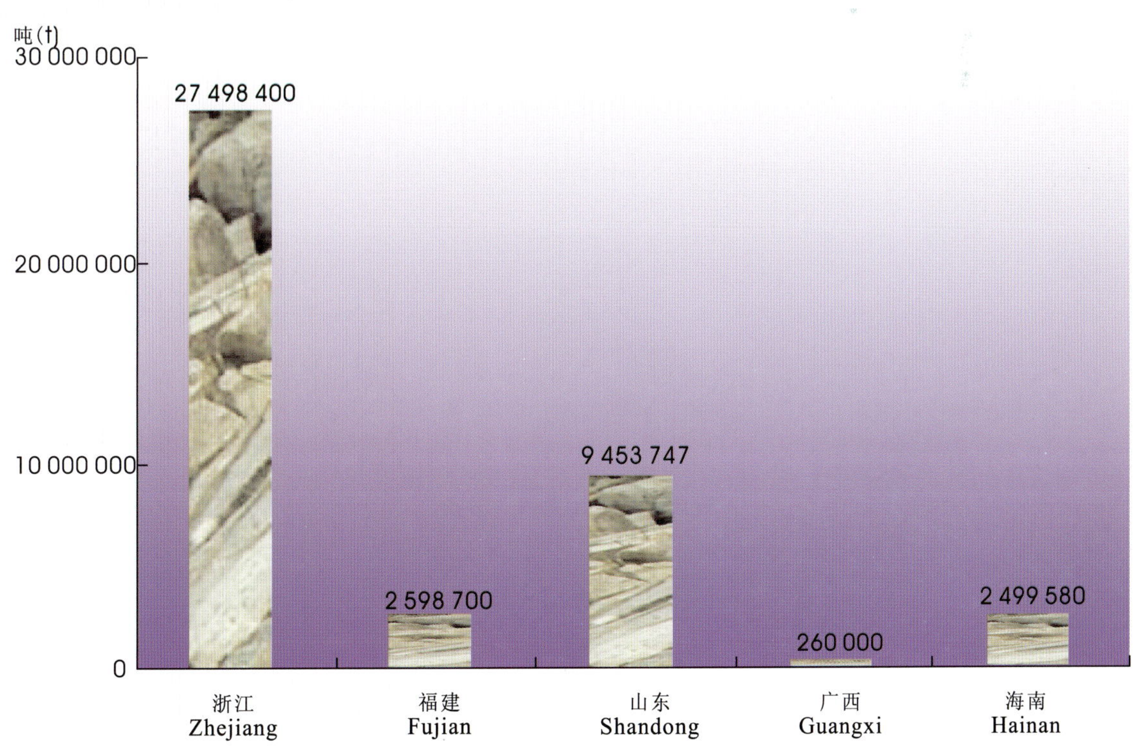

图9 2011年沿海地区海盐产量

Sea Salt Production by Coastal Regions in 2011

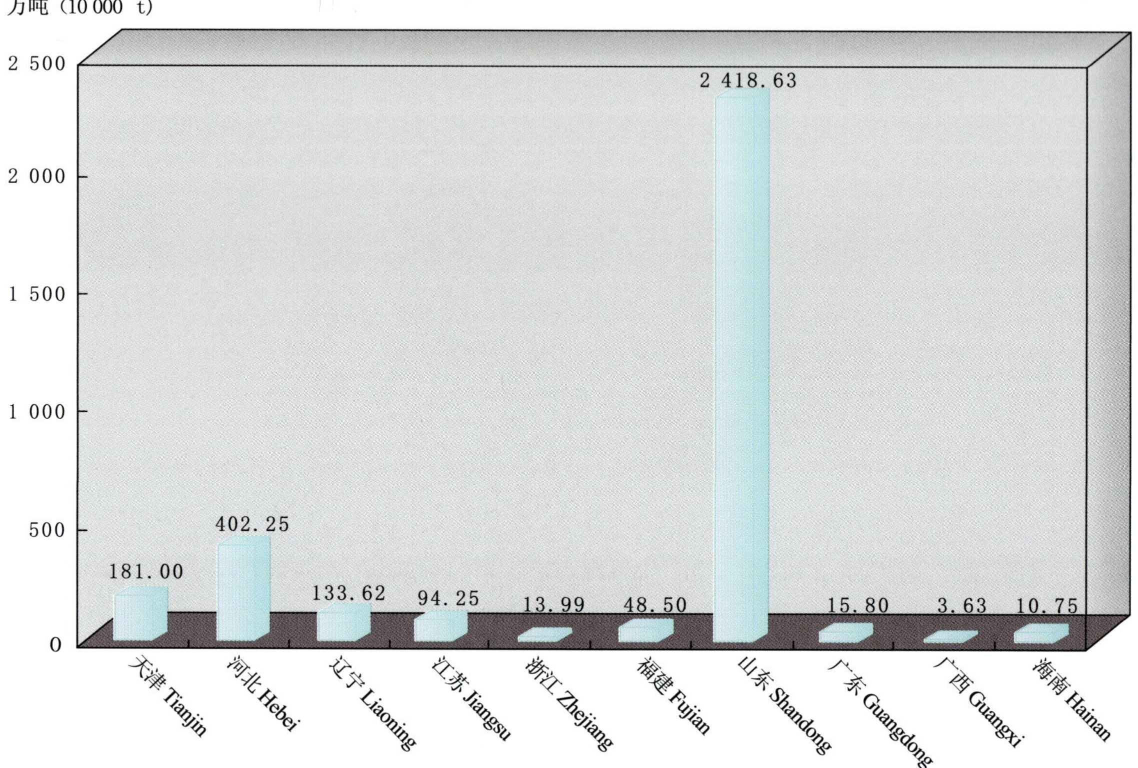

图10 2011年沿海地区海洋造船完工量

Completed Quantity of Marine Shipbuilding in the Coastal Region in 2011

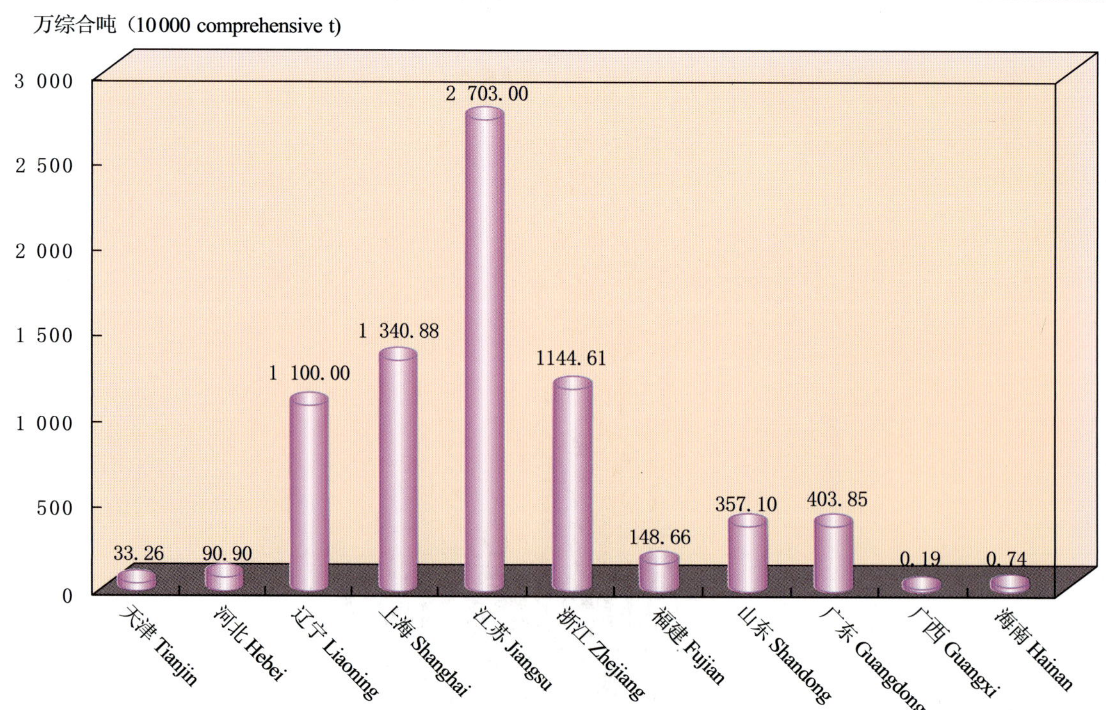

图11 2011年沿海地区海洋货物周转量

Goods Turnover Volume by Coastal Regions in 2011

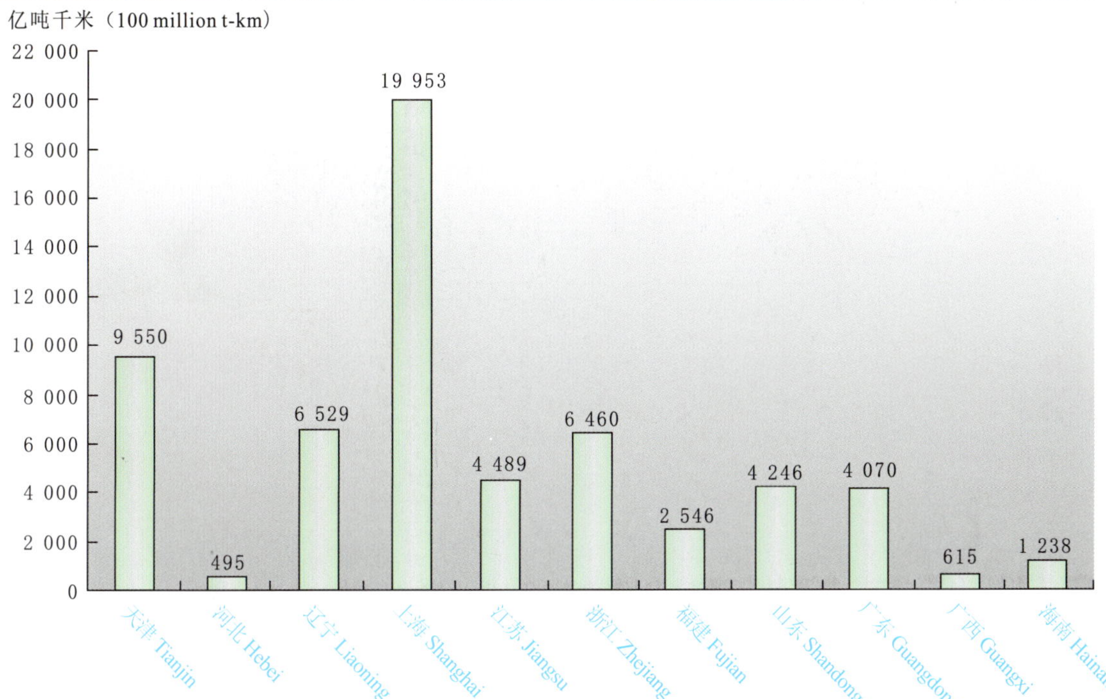

图12 2011年沿海港口国际标准集装箱吞吐量

International Standardized Containers Handled Coastal Seaports in 2011

图13 主要沿海城市国际旅游（外汇）收入
Foreign Exchange Earnings from International Tourism by Coastal Cities

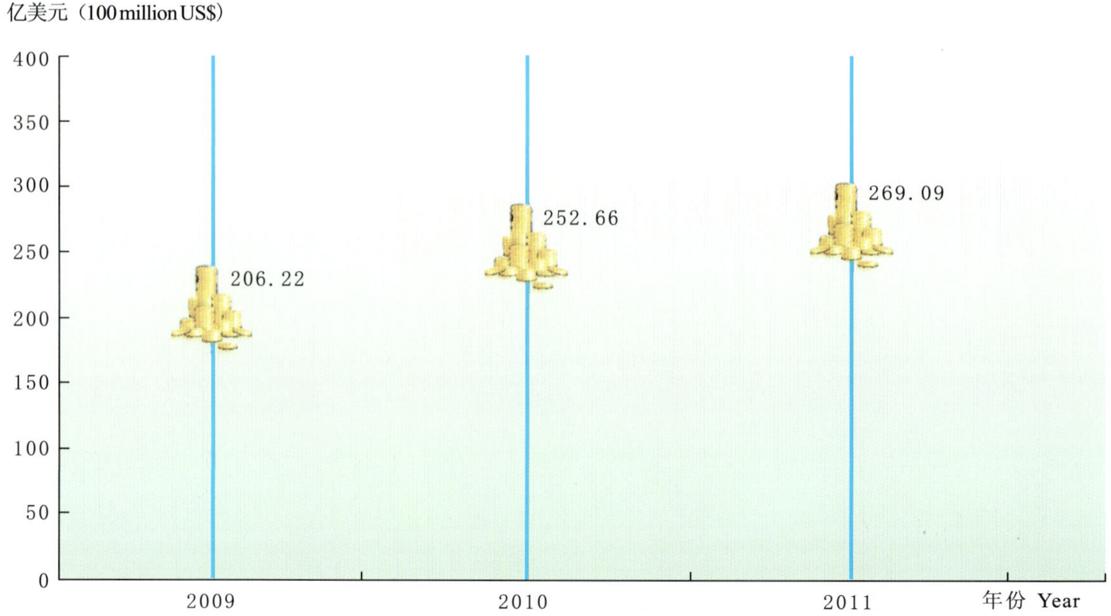

图14 主要沿海城市接待入境旅游者人数
Number of Oversea Visitor Arrivals in Major Coastal Cities

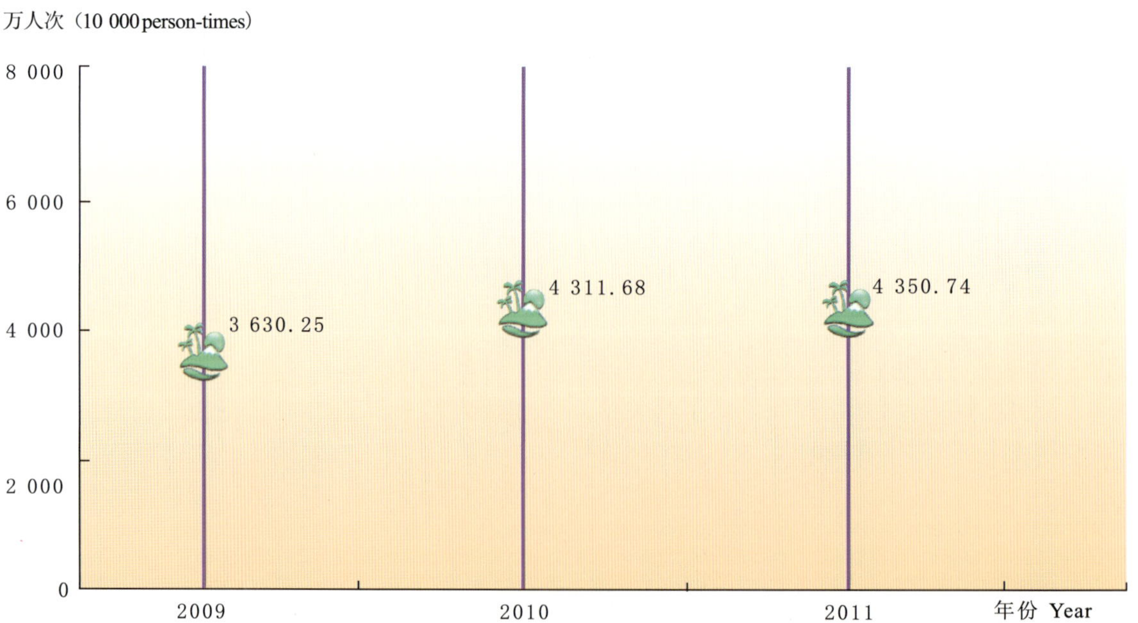

图15 2011年三大经济区地区生产总值与海洋生产总值
GDP and GOP in the Three Major Economic Zones in 2011

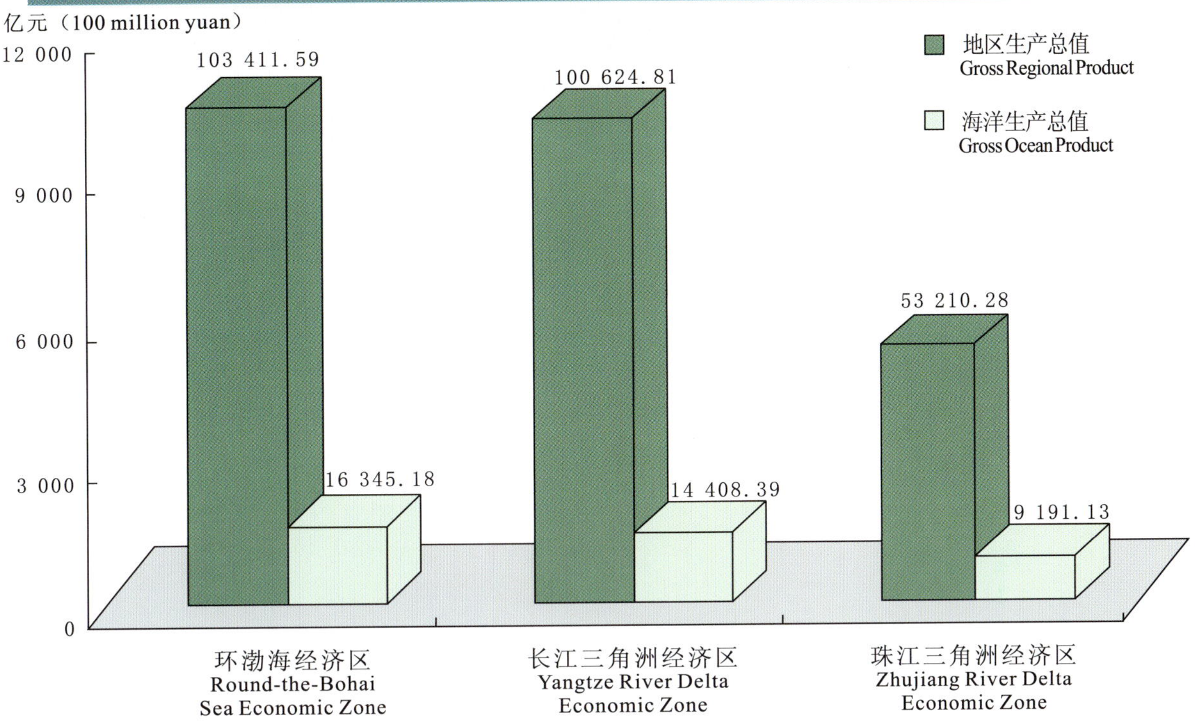

图16 2011年三大经济区海洋生产总值占全国海洋生产总值比重
Proportion of the GOP of the Three Major Economic Zones in the National GOP in 2011

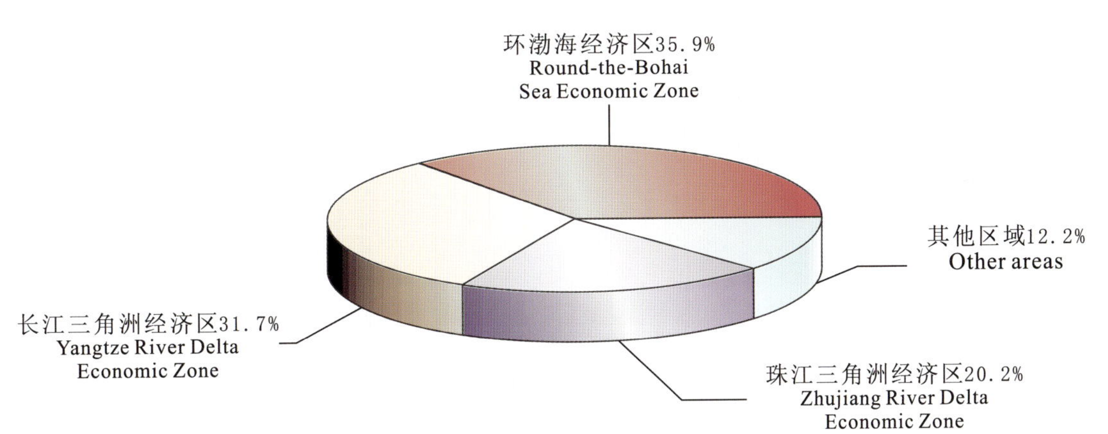

图17 2011年沿海地区海洋生产总值

Gross Ocean Product by Coastal Regions in 2011

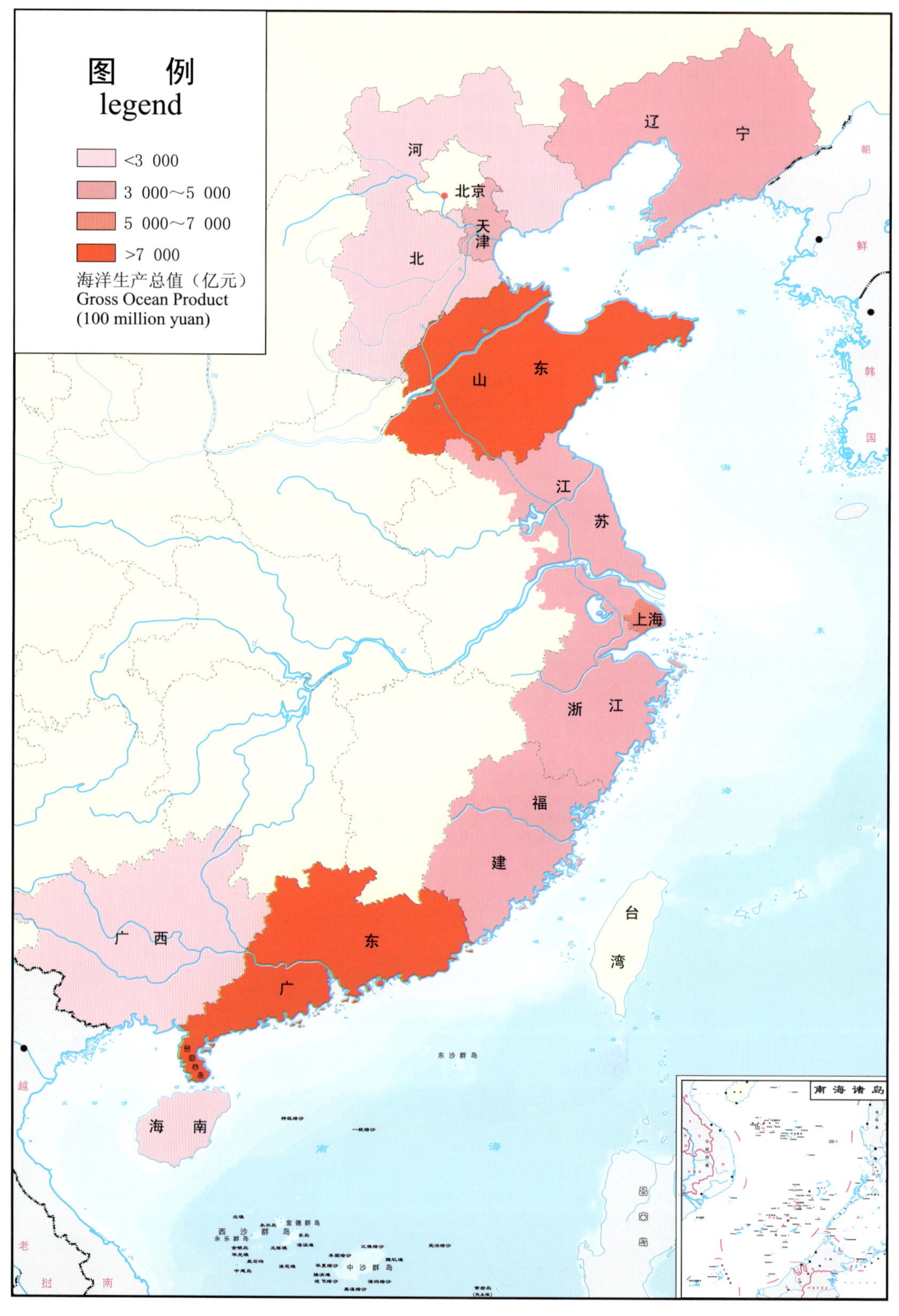

图18 2011年沿海地区海洋经济贡献

Marine Economic Contributions by Coastal Regions in 2011

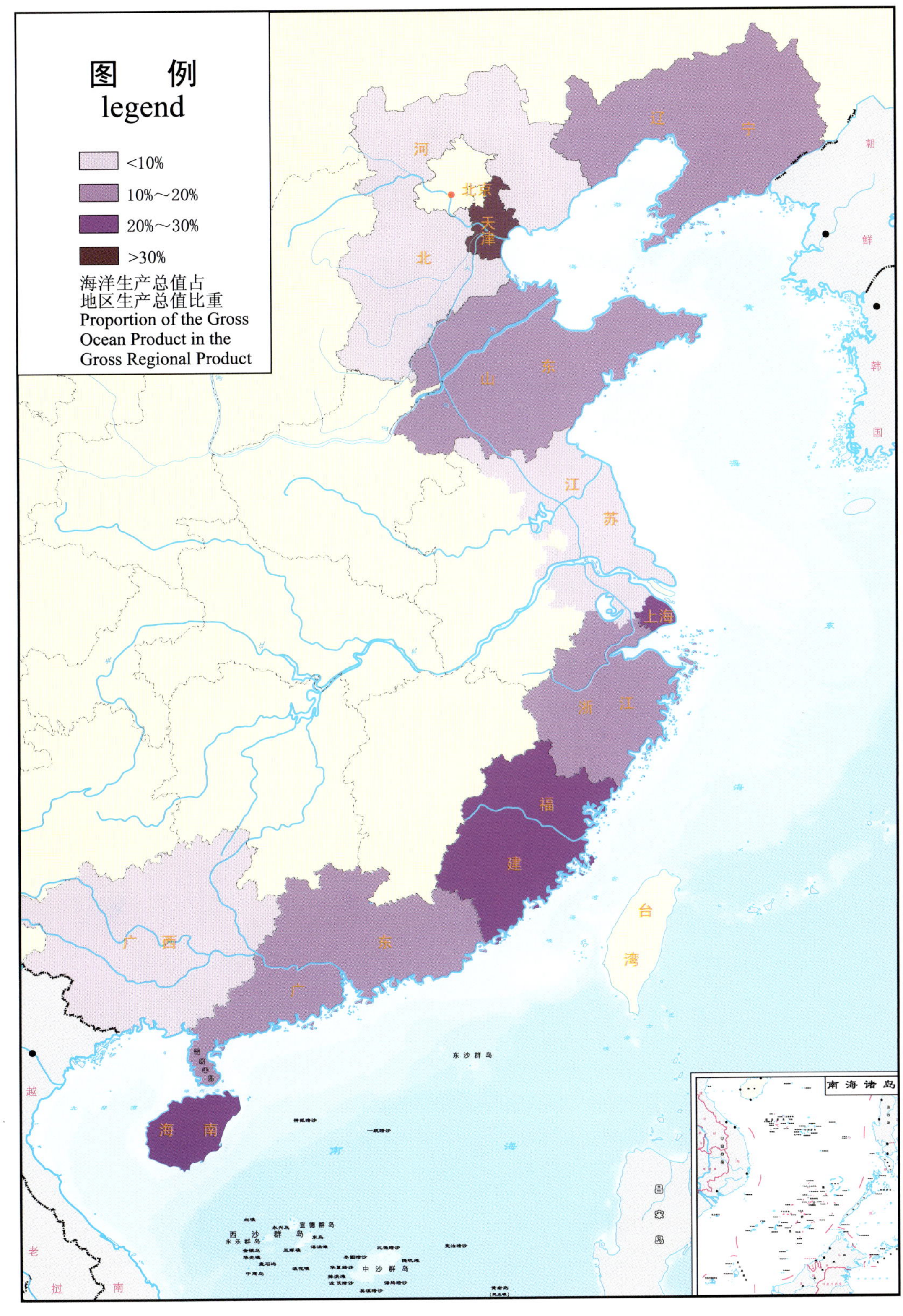

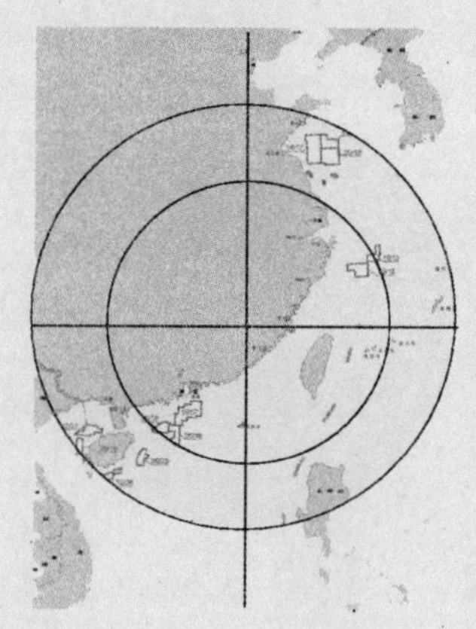

1

综 合 资 料

Integrated Data

1-1　沿海地区行政区划
Administrative Division of Coastal Regions

单位：个 (number)

沿海地区 Coastal Region	沿海城市 Coastal City	沿海地带 Coastal County (District)			
		合　计 Total	县 County	县级市 County-level city	区 District
合　计 Total	**53**	**237**	**64**	**56**	**117**
天　津 Tianjin	1	1			1
河　北 Hebei	3	11	6	1	4
辽　宁 Liaoning	6	22	4	7	11
上　海 Shanghai	1	5	1		4
江　苏 Jiangsu	3	15	8	4	3
浙　江 Zhejiang	7	35	11	10	14
福　建 Fujian	6	34	11	8	15
山　东 Shandong	7	37	6	14	17
广　东 Guangdong	14	56	11	6	39
广　西 Guangxi	3	8	1	1	6
海　南 Hainan	2	13	5	5	3

注：沿海地带中未包括广东省的东莞、中山和海南省的三亚。

Note: Dongguan, Zhongshan of Guangdong Province and Sanya of Hainan Province are not included in the Coastal County.

1-2 沿海行政区划一览表
Table of Administrative Division of Coastal Regions

沿海地区 Coastal Region	地区代码 Zip Code	沿海城市 Coastal City	地区代码 Zip Code	沿海地带 Coastal County (District)	地区代码 Zip Code
天 津 Tianjin	120000			滨海新区Binhai Xinqu	120116
河 北 Hebei	130000	唐山 Tangshan	130200	丰南区Fengnan Qu	130207
				滦南县Luannan Xian	130224
				乐亭县Leting Xian	130225
				唐海县Tanghai Xian	130230
		秦皇岛 Qinhuangdao	130300	海港区Haigang Qu	130302
				山海关区Shanhaiguan Qu	130303
				北戴河区Beidaihe Qu	130304
				昌黎县Changli Xian	130322
				抚宁县Funing Xian	130323
		沧州 Cangzhou	130900	海兴县Haixing Xian	130924
				黄骅市Huanghua Shi	130983
辽 宁 Liaoning	210000	大连 Dalian	210200	中山区Zhongshan Qu	210202
				西岗区Xigang Qu	210203
				沙河口区Shahekou Qu	210204
				甘井子区Ganjingzi Qu	210211
				旅顺口区Lüshunkou Qu	210212
				金州区Jinzhou Qu	210213
				长海县Changhai Xian	210224
				瓦房店市Wafangdian Shi	210281
				普兰店市Pulandian Shi	210282
				庄河市Zhuanghe Shi	210283
		丹东 Dandong	210600	东港市Donggang Shi	210681
		锦州 Jinzhou	210700	凌海市Linghai Shi	210781
		营口 Yingkou	210800	西市区Xishi Qu	210803
				鲅鱼圈区Bayuquan Qu	210804
				老边区Laobian Qu	210811
				盖州市Gaizhou Shi	210881
		盘锦 Panjin	211100	大洼县Dawa Xian	211121
				盘山县Panshan Xian	211122

1-2 续表1 continued

沿海地区 Coastal Region	地区代码 Zip Code	沿海城市 Coastal City	地区代码 Zip Code	沿海地带 Coastal County (District)	地区代码 Zip Code
		葫芦岛 Huludao	211400	连山区Lianshan Qu	211402
				龙港区Longgang Qu	211403
				绥中县Suizhong Xian	211421
				兴城市Xingcheng Shi	211481
上海 Shanghai	310000			宝山区Baoshan Qu	310113
				浦东新区Pudong Xinqu	310115
				金山区Jinshan Qu	310116
				奉贤区Fengxian Qu	310120
				崇明县Chongming Xian	310230
江苏 Jiangsu	320000	南通 Nantong	320600	通州区Tongzhou Qu	320612
				海安县Hai'an Xian	320621
				如东县Rudong Xian	320623
				启东市Qidong Shi	320681
				海门市Haimen Shi	320684
		连云港 Lianyungang	320700	连云区Lianyun Qu	320703
				新浦区Xinpu Qu	320705
				赣榆县Ganyu Xian	320721
				灌云县Guanyun Xian	320723
				灌南县Guannan Xian	320724
		盐城 Yancheng	320900	响水县Xiangshui Xian	320921
				滨海县Binhai Xian	320922
				射阳县Sheyang Xian	320924
				东台市Dongtai Shi	320981
				大丰市Dafeng Shi	320982
浙江 Zhejiang	330000	杭州 Hangzhou	330100	滨江区Binjiang Qu	330108
				萧山区Xiaoshan Qu	330109
		宁波 Ningbo	330200	海曙区Haishu Qu	330203
				江东区Jiangdong Qu	330204
				江北区Jiangbei Qu	330205
				北仑区Beilun Qu	330206
				镇海区Zhenhai Qu	330211
				鄞州区Yinzhou Qu	330212
				象山县Xiangshan Xian	330225
				宁海县Ninghai Xian	330226
				余姚市Yuyao Shi	330281
				慈溪市Cixi Shi	330282
				奉化市Fenghua Shi	330283

1-2 续表2 continued

沿海地区 Coastal Region	地区代码 Zip Code	沿海城市 Coastal City	地区代码 Zip Code	沿海地带 Coastal County (District)	地区代码 Zip Code
		温州 Wenzhou	330300	龙湾区Longwan Qu	330303
				瓯海区Ouhai Qu	330304
				洞头县Dongtou Xian	330322
				平阳县Pingyang Xian	330326
				苍南县Cangnan Xian	330327
				瑞安市Rui'an Shi	330381
				乐清市Yueqing Shi	330382
		嘉兴 Jiaxing	330400	海盐县Haiyan Xian	330424
				海宁市Haining Shi	330481
				平湖市Pinghu Shi	330482
		绍兴 Shaoxing	330600	绍兴县Shaoxing Xian	330621
				上虞市Shangyu Shi	330682
		舟山 Zhoushan	330900	定海区Dinghai Qu	330902
				普陀区Putuo Qu	330903
				岱山县Daishan Xian	330921
				嵊泗县Shengsi Xian	330922
		台州 Taizhou	331000	椒江区Jiaojiang Qu	331002
				路桥区Luqiao Qu	331004
				玉环县Yuhuan Xian	331021
				三门县Sanmen Xian	331022
				温岭市Wenling Shi	331081
				临海市Linhai Shi	331082
福建 Fujian	350000	福州 Fuzhou	350100	马尾区Mawei Qu	350105
				连江县Lianjiang Xian	350122
				罗源县Luoyuan Xian	350123
				平潭县Pingtan Xian	350128
				福清市Fuqing Shi	350181
				长乐市Changle Shi	350182
		厦门 Xiamen	350200	思明区Siming Qu	350203
				海沧区Haicang Qu	350205

1-2 续表3 continued

沿海地区 Coastal Region	地区代码 Zip Code	沿海城市 Coastal City	地区代码 Zip Code	沿海地带 Coastal County (District)	地区代码 Zip Code
				湖里区Huli Qu	350206
				集美区Jimei Qu	350211
				同安区Tong'an Qu	350212
				翔安区Xiang'an Qu	350213
		莆田 Putian	350300	城厢区Chengxiang Qu	350302
				涵江区Hanjiang Qu	350303
				荔城区Licheng Qu	350304
				秀屿区Xiuyu Qu	350305
				仙游县Xianyou Xian	350322
		泉州 Quanzhou	350500	丰泽区Fengze Qu	350503
				洛江区Luojiang Qu	350504
				泉港区Quangang Qu	350505
				惠安县Hui'an Xian	350521
				金门县Jinmen Xian	350527
				石狮市Shishi Shi	350581
				晋江市Jinjiang Shi	350582
				南安市Nan'an Shi	350583
		漳州 Zhangzhou	350600	云霄县Yunxiao Xian	350622
				漳浦县Zhangpu Xian	350623
				诏安县Zhao'an Xian	350624
				东山县Dongshan Xian	350626
				龙海市Longhai Shi	350681
		宁德 Ningde	350900	蕉城区Jiaocheng Qu	350902
				霞浦县Xiapu Xian	350921
				福安市Fu'an Shi	350981
				福鼎市Fuding Shi	350982
山东 Shandong	370000	青岛 Qingdao	370200	市南区Shinan Qu	370202
				市北区Shibei Qu	370203
				四方区Sifang Qu	370205
				黄岛区Huangdao Qu	370211
				崂山区Laoshan Qu	370212
				李沧区Licang Qu	370213
				城阳区Chengyang Qu	370214
				胶州市Jiaozhou Shi	370281
				即墨市Jimo Shi	370282
				胶南市Jiaonan Shi	370284

1-2 续表4 continued

沿海地区 Coastal Region	地区代码 Zip Code	沿海城市 Coastal City	地区代码 Zip Code	沿海地带 Coastal County (District)	地区代码 Zip Code
		东营 Dongying	370500	东营区Dongying Qu	370502
				河口区Hekou Qu	370503
				垦利县Kenli Xian	370521
				利津县Lijin Xian	370522
				广饶县Guangrao Xian	370523
		烟台 Yantai	370600	芝罘区Zhifu Qu	370602
				福山区Fushan Qu	370611
				牟平区Muping Qu	370612
				莱山区Laishan Qu	370613
				长岛县Changdao Xian	370634
				龙口市Longkou Shi	370681
				莱阳市Laiyang Shi	370682
				莱州市Laizhou Shi	370683
				蓬莱市Penglai Shi	370684
				招远市Zhaoyuan Shi	370685
				海阳市Haiyang Shi	370687
		潍坊 Weifang	370700	寒亭区Hanting Qu	370703
				寿光市Shouguang Shi	370783
				昌邑市Changyi Shi	370786
		威海 Weihai	371000	环翠区Huancui Qu	371002
				文登市Wendeng Shi	371081
				荣成市Rongcheng Shi	371082
				乳山市Rushan Shi	371083
		日照 Rizhao	371100	东港区Donggang Qu	371102
				岚山区Lanshan Qu	371103
		滨州 Binzhou	371600	无棣县Wudi Xian	371623
				沾化县Zhanhua Xian	371624
广东 Guangdong	440000	广州 Guangzhou	440100	荔湾区Liwan Qu	440103
				越秀区Yuexiu Qu	440104
				海珠区Haizhu Qu	440105
				天河区Tianhe Qu	440106
				白云区Baiyun Qu	440111
				黄埔区Huangpu Qu	440112
				番禺区Panyu Qu	440113
				南沙区Nansha Qu	440115
				萝岗区Luogang Qu	440116

1-2 续表5 continued

沿海地区 Coastal Region	地区代码 Zip Code	沿海城市 Coastal City	地区代码 Zip Code	沿海地带 Coastal County (District)	地区代码 Zip Code
		深圳 Shenzhen	440300	罗湖区Luohu Qu	440303
				福田区Futian Qu	440304
				南山区Nanshan Qu	440305
				宝安区Bao'an Qu	440306
				龙岗区Longgang Qu	440307
				盐田区Yantian Qu	440308
		珠海 Zhuhai	440400	香洲区Xiangzhou Qu	440402
				斗门区Doumen Qu	440403
				金湾区Jinwan Qu	440404
		汕头 Shantou	440500	龙湖区Longhu Qu	440507
				金平区Jinping Qu	440511
				濠江区Haojiang Qu	440512
				潮阳区Chaoyang Qu	440513
				潮南区Chaonan Qu	440514
				澄海区Chenghai Qu	440583
				南澳县Nan'ao Xian	440523
		江门 Jiangmen	440700	蓬江区Pengjiang Qu	440703
				江海区Jianghai Qu	440704
				新会区Xinhui Qu	440705
				台山市Taishan Shi	440781
				恩平市Enping Shi	440785
		湛江 Zhanjiang	440800	赤坎区Chikan Qu	440802
				霞山区Xiashan Qu	440803
				坡头区Potou Qu	440804
				麻章区Mazhang Qu	440811
				遂溪县Suixi Xian	440823
				徐闻县Xuwen Xian	440825
				廉江市Lianjiang Shi	440881
				雷州市Leizhou Shi	440882
				吴川市Wuchuan Shi	440883
		茂名 Maoming	440900	茂南区Maonan Qu	440902
				茂港区Maogang Qu	440903
				电白县Dianbai Xian	440923
		惠州 Huizhou	441300	惠城区Huicheng Qu	441302
				惠阳区Huiyang Qu	441303
				惠东县Huidong Xian	441323

1-2 续表6 continued

沿海地区 Coastal Region	地区代码 Zip Code	沿海城市 Coastal City	地区代码 Zip Code	沿海地带 Coastal County (District)	地区代码 Zip Code
		汕尾 Shanwei	441500	城　区Chengqu	441502
				海丰县Haifeng Xian	441521
				陆丰市Lufeng Shi	441581
		阳江 Yangjiang	441700	江城区Jiangcheng Qu	441702
				阳西县Yangxi Xian	441721
				阳东县Yangdong Xian	441723
		东莞 Dongguan	441900		
		中山 Zhongshan	442000		
		潮州 Chaozhou	445100	湘桥区Xiangqiao Qu	445102
				饶平县Raoping Xian	445122
		揭阳 Jieyang	445200	榕城区Rongcheng Qu	445202
				揭东县Jiedong Xian	445221
				惠来县Huilai Xian	445224
广西 Guangxi	450000	北海 Beihai	450500	海城区Haicheng Qu	450502
				银海区Yinhai Qu	450503
				铁山港区Tieshangang Qu	450512
				合浦县Hepu Xian	450521
		防城港 Fangchenggang	450600	港口区Gangkou Qu	450602
				防城区Fangcheng Qu	450603
				东兴市Dongxing Shi	450681
		钦州 Qinzhou	450700	钦南区Qinnan Qu	450702
海南 Hainan	460000	海口 Haikou	460100	秀英区Xiuying Qu	460105
				龙华区Longhua Qu	460106
				美兰区Meilan Qu	460108
		三亚 Sanya	460200		
				琼海市 Qionghai Shi	469002
				儋州市 Danzhou Shi	469003
				文昌市 Wenchang Shi	469005
				万宁市 Wanning Shi	469006
				东方市 Dongfang Shi	469007
				澄迈县Chengmai Xian	469023
				临高县Lingao Xian	469024
				昌江黎族自治县 Changjiang Lizu Zizhixian	469026
				乐东黎族自治县 Ledong Lizu Zizhixian	469027
				陵水黎族自治县 Lingshui Lizu Zizhixian	469028

1-3 海洋自然地理
Marine Physical Geography

指　　标		Item	指标值 Data
海洋平均深度	（米）	Average Depth of Sea (m)	961
海洋最大深度	（米）	Maximum Depth of Sea (m)	5 559
岸线总长度	（千米）	Length of Coastline (km)	32 000
大陆岸线长度		Mainland Shore	18 000
岛屿岸线长度		Island Shore	14 000
＞500m^{2}岛屿个数	（个）	Number of Islands ＞500 m^2 each (unit)	7 300
岛屿面积	（万平方千米）	Area of Islands (10 000 km^2)	8
已利用的无居民海岛	（个）	Uninhabited Islands Which Have Been Utilized	1 900
特殊用途海岛		Islands for Special Purposes	1 020
公共服务用岛		Island Use for Public Service	365
旅游娱乐用岛		Island Use for Tourism and Recreation	73
农林牧渔业用岛		Island Use for Agriculture, Forestry, Animal Husbandry and Fishery	340
工业、仓储、交通运输用岛		Island User for Industry, Storage, Communications and Transport	49

注：海岛数据来源于《全国海岛保护规划》。

Note: Data on islands are derived from the National Plan for Island protection.

1-4 海区海洋石油储量
Offshore Oil Reserves in the Sea Area

自然海区名称 Natural Sea Area	海洋石油（万吨） Offshore Oil (10 000 t)	
	累计探明技术可采储量 Proven Technically Recoverable Reserves in the Aggregate	剩余技术可采储量 Surplus Technically Recoverable Reserves
合 计 Total	**87 464.0**	**45 043.4**
渤 海 Bohai Sea	50 465.6	31 926.6
黄 海 Huanghai Sea		
东 海 Donghai Sea	1 297.3	891.1
南 海 Nanhai Sea	35 701.0	12 225.8

注：数据来源于《2011年全国矿产资源储量通报》。

Note: The data come from the *Journal on the National Mineral Resources Reserves in 2011* .

1-5 沿海地区水资源情况
Water Resources by Coastal Regions

地 区 Region	水资源总量（亿立方米） Total Amount of Water Resources (100 million m^3)	地表水资源量 Surface Water Resources	地下水资源量 Groundwater Resources	地表水与地下水资源重复量 Duplicated Measurement Between Surface Water and Groundwater	人均水资源量（立方米/人） Per Capita Water Resources (m^3/person)
全国总计 National Total	**23 258.5**	**22 215.2**	**7 214.8**	**6 171.5**	**1 730.4**
天 津 Tianjin	15.4	10.9	5.2	0.7	116.0
河 北 Hebei	157.2	69.8	126.2	38.9	217.7
辽 宁 Liaoning	294.8	260.5	111.9	77.6	673.2
上 海 Shanghai	20.7	16.2	7.4	2.9	89.1
江 苏 Jiangsu	492.4	399.0	115.1	21.7	624.6
浙 江 Zhejiang	745.0	733.3	184.2	172.5	1 365.7
福 建 Fujian	774.9	773.5	243.4	242.1	2 090.5
山 东 Shandong	347.6	237.5	195.9	85.8	361.6
广 东 Guangdong	1 471.3	1 461.3	362.1	352.1	1 404.8
广 西 Guangxi	1 350.0	1 350.0	271.2	271.2	2 917.4
海 南 Hainan	484.1	478.8	111.6	106.3	5 545.6

注： 数据来源于《2012中国统计年鉴》。

Note: The data come from the *China Statistical Yearbook 2012*.

1-6 沿海地区湿地面积
Area of Wetlands by Coastal Regions

地　区 Region	湿地面积 （千公顷） Area of Wetlands (1 000 hm^2)	近岸及海岸 Coasts and Seashores	湿地面积 占国土面积 比重（%） Proportion of Wetlands in Total Area of Territory (%)
全国总计 National Total	**38 485.5**	**5 941.7**	**4.01**
天 津 Tianjin	171.8	58.1	14.95
河 北 Hebei	1 081.9	278.8	5.82
辽 宁 Liaoning	1 219.6	738.1	8.37
上 海 Shanghai	319.7	305.4	53.68
江 苏 Jiangsu	1 674.7	843.5	16.32
浙 江 Zhejiang	802.2	574.3	7.88
福 建 Fujian	443.0	370.6	3.65
山 东 Shandong	1 784.1	1 210.9	11.72
广 东 Guangdong	1 398.1	1 017.8	7.86
广 西 Guangxi	656.1	348.4	2.76
海 南 Hainan	311.5	190.0	9.13

注：数据来源于《2012中国统计年鉴》。

Note: The data come from the *China Statistical Yearbook 2012*.

1-7 红树林各地类面积
Site Classification and Area of Sharpleaf Mangrove (*Rhizophora Apiculata*)

单位：公顷 (hm²)

地 区 Region	红树林各地类总面积 Total Site Area of Sharpleaf Mangrove	现有面积 Established	未成林面积 Unestablished	宜林地面积 Suitable for Planting
全国总计 National Total	**82 757.2**	**22 024.9**	**1 884.1**	**58 848.2**
浙 江 Zhejiang	5 452.3	20.6	236.1	5 195.6
福 建 Fujian	13 410.1	615.1	286.4	12 508.6
广 东 Guangdong	32 325.9	9 084.0	981.3	22 260.6
广 西 Guangxi	18 029.2	8 374.9	380.3	9 274.0
海 南 Hainan	13 539.7	3 930.3		9 609.4

注：数据来源于《2012中国统计年鉴》。

Note: The data come from the *China Statistical Yearbook 2012*.

1-8　主要沿海城市气候基本情况
Climate of Major Coastal Cities

城 市 City	年平均气温 （摄氏度） Annual Average Temperature(℃)	年平均相对湿度 （%） Annual Average Relative Humidity (%)	全年降水量 （毫米） Annual Precipitation (mm)	全年日照时数 （小时） Annual Sunshine Hours (h)
天 津 Tianjin	12.9	54	485.8	2 245.2
上 海 Shanghai	16.9	69	1 009.1	1 663.7
杭 州 Hangzhou	17.2	69	1 359.9	1 496.1
福 州 Fuzhou	20.2	70	1 244.9	1 394.8
广 州 Guangzhou	21.4	74	1 632.3	1 878.4
海 口 Haikou	23.3	81	2 002.1	1 486.2

注：数据来源于《2012中国统计年鉴》。

Note: The data come from the *China Statistical Yearbook 2012*.

主要统计指标解释

1. 沿海地区　即广义的沿海地区，是指有海岸线（大陆岸线和岛屿岸线）的地区，按行政区划分为沿海省、自治区、直辖市。

2. 沿海城市　是指有海岸线的直辖市和地级市（包括其下属的全部区、县和县级市）。

3. 沿海地带　即狭义的沿海地区，是指有海岸线的县、县级市、区（包括直辖市和地级市的区）。

4. 海洋　是海和洋的统称。洋为地球表面上相连接的广大咸水水体的主体部分。海为地球表面相连接的广大咸水水体被陆地、岛礁、半岛包围或分隔的边缘部分。

5. 水资源总量　指评价区内降水形成的地表和地下产水总量，即地表产流量与降水入渗补给地下水量之和，不包括过境水量。

6. 地表水资源量　指评价区内河流、湖泊、冰川等地表水体中可以逐年更新的动态水量，即当地天然河川径流量。

7. 地下水资源量　指评价区内降水和地表水对饱水岩土层的补给量，包括降水入渗补给量和河道、湖库、渠系、渠灌田间等地表水体的入渗补给量。

8. 地表水与地下水资源重复量　指地表水和地下水相互转化的部分，即天然河川径流量中的地下水排泄量和地下水补给量中来源于地表水的入渗补给量。

9. 湿地　指天然或人工、长久或暂时性的沼泽地、泥炭地或水域地带，包括静止或流动、淡水、半咸水、咸水体，低潮时水深不超过6米的水域以及海岸地带地区的珊瑚滩和海草床、滩涂、红树林、河口、河流、淡水沼泽、沼泽森林、湖泊、盐沼及盐湖。

10. 红树林　指生长在热带、亚热带低能海岸潮间带上部，受周期性潮水浸淹，以红树植物为主体的常绿灌木或乔木组成的潮滩湿地木本生物群落。

11. 气温　指空气的温度，我国一般以摄氏度(℃)为单位表示。气象观测的温度表是放在离地面约1.5米处通风良好的百叶箱里测量的，因此，通常说的气温指的是离地面1.5米处百叶箱中的温度。其统计计算方法为：

　　月平均气温是将全月各日的平均气温相加，除以该月的天数而得。

　　年平均气温是将12个月的月平均气温累加后除以12而得。

12. 相对湿度　指空气中实际所含水蒸气密度和同温度下饱和水蒸气密度的百分比值。其统计方法与气温相同。

13. 降水量　指从天空降落到地面的液态或固态(经融化后)水，未经蒸发、渗透、流失而在地面上积聚的深度。其统计计算方法为：

　　月降水量是将全月各日的降水量累加而得。

　　年降水量是将12个月的月降水量累加而得。

14. 日照时数　指太阳实际照射地面的时间。其统计方法与降水量相同。

Explanatory Notes on Main Statistical Indicators

1. Coastal Region, i.e., the coastal region in a broad sense, refers to the regions with coastlines (continental and island coastlines), which are divided into the coastal provinces, autonomous regions and municipalities directly under the Central Government according to the administrative zoning.

2. Coastal City refers to the municipalities directly under the Central Government and the prefecture-level cities (including all the districts, counties and county-level cities under them).

3. Coastal Zone, i.e., the coastal region in a narrow sense, refers to the counties, county-level cities and districts with coastlines (including the districts under the municipalities directly under the Central Government and the prefecture-level districts).

4. Ocean is the general name for sea and ocean. Ocean refers to the main body of large salt water connected with the earth surface. Sea refers to the edge areas of the salt water on the earth surface that are compartmentalized or surrounded by land, island, reef or peninsula.

5. Total Water Resources refers to total volume of water resources measured as run-off for surface water from rainfall and recharge for groundwater in a given area, excluding transit water.

6. Surface Water Resources refers to total renewable resources which exist in rivers, lakes, glaciers and other collectors from rainfall and are measured as run-off of rivers.

7. Groundwater Resources refers to replenishment of aquifers with rainfall and surface water.

8. Duplicated Measurement between Surface Water and Groundwater refers to the exchange between surface water and groundwater, i.e. run-off of rivers includes some depletion into groundwater while groundwater includes some replenishment from surface water.

9. Wetlands refer to marshland and peat bog, whether natural or man-made, permanent or temporary; water covered areas, whether stagnant or flowing, with fresh or brackish-fresh or salty water that is less than 6 meters deep at low tide; as well as coral beach, weed beach, mud beach, mangrove, river outlet, rivers, fresh-water marshland, marshland forests, lakes, salty bog and salt lakes along the coastal areas.

10. Mangrove refers to evergreen woody plants or plant communities in tropical or sub-tropical zones which live between the sea and the land in areas which are inundated by tides.

11. Temperature refers to the air temperature. China uses centigrade as the unit. The thermometry used for weather observation is put in a breezy shutter, which is 1.5 meters high from the ground. Therefore, the commonly used temperature refers to the temperature in the breezy shutter 1.5 meters away from the ground. The calculation method is as follows:

Monthly average temperature is the summation of average daily temperature of one month divided by the actual days of that particular month.

Annual average temperature is the summation of monthly averages of a year divided by 12 months.

12. Relative Humidity refers to the ratio of actual water vapour pressure to the saturated water vapour density under the current temperature. The calculation method is the same as that of temperature.

13. Volume of Precipitation refers to the deepness of liquid state or solid state (thawed) water falling from the sky to the ground that has not evaporated, infiltrated or run off. The calculation method is as follows:

Monthly precipitation is the summation of daily precipitation of a month.

Annual precipitation is the summation of 12 months precipitation of a year.

14. Sunshine Hours refer to the actual hours of sun irradiating the earth. The calculation method is the same as that of the precipitation.

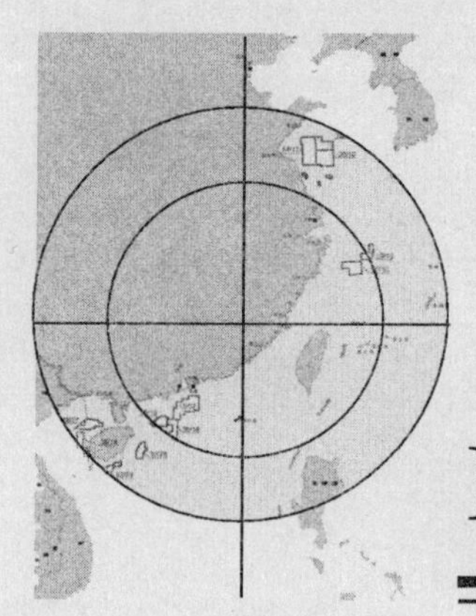

2 海洋经济核算

Marine Economic Accounting

2-1 全国海洋生产总值
National Gross Ocean Product

年 份 Year	海洋生产总值（亿元） Gross Ocean Product (100 million yuan)	第一产业 Primary Industry	第二产业 Secondary Industry	第三产业 Tertiary Industry	海洋生产总值占国内生产总值比重（%） Proportion of the Gross Ocean Product in GDP (%)	海洋生产总值增长速度（%） Growth Rate of the Gross Ocean Product (%)
2001	9 518.4	646.3	4 152.1	4 720.1	8.68	
2002	11 270.5	730.0	4 866.2	5 674.3	9.37	19.8
2003	11 952.3	766.2	5 367.6	5 818.5	8.80	4.2
2004	14 662.0	851.0	6 662.8	7 148.2	9.17	16.9
2005	17 655.6	1 008.9	8 046.9	8 599.8	9.55	16.3
2006	21 592.4	1 228.8	10 217.8	10 145.7	9.98	18.0
2007	25 618.7	1 395.4	12 011.0	12 212.3	9.64	14.8
2008	29 718.0	1 694.3	13 735.3	14 288.4	9.46	9.9
2009	32 277.6	1 857.7	14 980.3	15 439.5	9.47	9.2
2010	39 572.7	2 008.0	18 935.0	18 629.8	9.86	14.7
2011	45 496.0	2 381.9	21 685.6	21 428.5	9.62	9.9

2-2 全国海洋生产总值构成
Composition of National Gross Ocean Product

单位：% (%)

年 份 Year	第一产业 Primary Industry	第二产业 Secondary Industry	第三产业 Tertiary Industry
2001	6.8	43.6	49.6
2002	6.5	43.2	50.3
2003	6.4	44.9	48.7
2004	5.8	45.4	48.8
2005	5.7	45.6	48.7
2006	5.7	47.3	47.0
2007	5.4	46.9	47.7
2008	5.7	46.2	48.1
2009	5.8	46.4	47.8
2010	5.1	47.8	47.1
2011	5.2	47.7	47.1

2-3 海洋及相关产业增加值
Added Values of Marine and Related Industries

单位：亿元 (100 million yuan)

年 份 Year	合 计 Total	海洋产业 Marine Industry	主要海洋产业 Major Marine Industry	海洋科研教育管理服务业 Industries of Marine Scientific Research, Education, Management and Service	海洋相关产业 Ocean-related Industries
2001	9 518.4	5 733.6	3 856.6	1 877.0	3 784.8
2002	11 270.5	6 787.3	4 696.8	2 090.5	4 483.2
2003	11 952.3	7 137.7	4 754.4	2 383.3	4 814.6
2004	14 662.0	8 710.1	5 827.7	2 882.5	5 951.9
2005	17 655.6	10 539.0	7 188.0	3 350.9	7 116.6
2006	21 592.4	12 696.7	8 790.4	3 906.4	8 895.6
2007	25 618.7	15 070.6	10 478.3	4 592.3	10 548.0
2008	29 718.0	17 591.2	12 176.0	5 415.2	12 126.8
2009	32 277.6	18 822.0	12 843.6	5 978.4	13 455.6
2010	39 572.7	22 831.0	16 187.8	6 643.1	16 741.7
2011	45 496.0	26 422.0	18 865.2	7 556.8	19 074.1

2-4 海洋及相关产业增加值构成
Composition of the Added Value of Marine and Related Industries

单位：% (%)

年 份 Year	合 计 Total	海洋产业 Marine Industry	主要海洋产业 Major Marine Industry	海洋科研教育管理服务业 Industries of Marine Scientific Research, Education, Management and Service	海洋相关产业 Ocean-related Industries
2001	100.0	60.2	40.5	19.7	39.8
2002	100.0	60.2	41.7	18.5	39.8
2003	100.0	59.7	39.8	19.9	40.3
2004	100.0	59.4	39.7	19.7	40.6
2005	100.0	59.7	40.7	19.0	40.3
2006	100.0	58.8	40.7	18.1	41.2
2007	100.0	58.8	40.9	17.9	41.2
2008	100.0	59.2	41.0	18.2	40.8
2009	100.0	58.3	39.8	18.5	41.7
2010	100.0	57.7	40.9	16.8	42.3
2011	100.0	58.1	41.5	16.6	41.9

2-5 全国主要海洋产业增加值
Gross Output Value and Added Value of Major Marine Industries

主要海洋产业 Major Marine Industry	增加值 (亿元) Added Value (100 million yuan)	比上年增长(%) (按可比价计算) Percentage of Increase Over Last Year (%) (at comparable price)
合计 **Total**	**18 865.2**	**9.9**
海洋渔业 Marine Fishery Industry	3 202.9	1.1
海洋油气业 Offshore Oil and Natural Gas Industry	1 719.7	6.0
海洋矿业 Marine Mining Industry	53.3	2.6
海洋盐业 Sea Salt Industry	76.8	7.4
海洋船舶工业 Marine Shipbuilding Industry	1 352.0	10.8
海洋化工业 Marine Chemical Industry	695.9	3.2
海洋生物医药业 Marine Biomedicine Industry	150.8	21.3
海洋工程建筑业 Marine Engineering Architecture Industry	1 086.8	13.9
海洋电力业 Marine Electric Power Industry	59.2	52.8
海水利用业 Marine Seawater Utilization Industry	10.4	11.9
海洋交通运输业 Maritime Communications and Transportation Industry	4 217.5	10.8
滨海旅游业 Coastal Tourism	6 239.9	12.1

2-6　海洋渔业增加值
Added Value of Marine Fishery Industry

单位：亿元　(100 million yuan)

年　份 Year	增加值 Added Value
2001	966.0
2002	1 091.2
2003	1 145.0
2004	1 271.2
2005	1 507.6
2006	1 672.0
2007	1 906.0
2008	2 228.6
2009	2 440.8
2010	2 851.6
2011	3 202.9

2-7　海洋油气业增加值
Added Value of Offshore Oil and Gas Industry

单位：亿元　(100 million yuan)

年　份 Year	增加值 Added Value
2001	176.8
2002	181.8
2003	257.0
2004	345.1
2005	528.2
2006	668.9
2007	666.9
2008	1 020.5
2009	614.1
2010	1 302.2
2011	1 719.7

2-8 海洋矿业增加值
Added Value of Marine Mining Industry

单位：亿元 (100 million yuan)

年 份 Year	增加值 Added Value
2001	1.0
2002	1.9
2003	3.1
2004	7.9
2005	8.3
2006	13.4
2007	16.3
2008	35.2
2009	41.6
2010	45.2
2011	53.3

注：自2008年起部分地区统计矿种增加。

Note: The data from 2008 include added kinds of minerals in some regions.

2-9 海洋盐业增加值
Added Value of Marine Salt Industry

单位：亿元 (100 million yuan)

年 份 Year	增加值 Added Value
2001	32.6
2002	34.2
2003	28.4
2004	39.0
2005	39.1
2006	37.1
2007	39.9
2008	43.6
2009	43.6
2010	65.5
2011	76.8

2-10 海洋船舶工业增加值
Added Value of Marine Shipbuilding Industry

单位：亿元 (100 million yuan)

年 份 Year	增加值 Added Value
2001	109.3
2002	117.4
2003	152.8
2004	204.1
2005	275.5
2006	339.5
2007	524.9
2008	742.6
2009	986.5
2010	1 215.6
2011	1 352.0

2-11 海洋化工业增加值
Added Value of Marine Chemical Industry

单位：亿元 (100 million yuan)

年 份 Year	增加值 Added Value
2001	64.7
2002	77.1
2003	96.3
2004	151.5
2005	153.3
2006	440.4
2007	506.6
2008	416.8
2009	465.3
2010	613.8
2011	695.9

注：自2006年起部分地区统计产品品种增加。

Note: The data from 2006 include added kinds of statistical products in some regions.

2-12　海洋生物医药业增加值
Added Value of Marine Biomedicine Industry

单位：亿元　　(100 million yuan)

年　份 Year	增加值 Added Value
2001	5.7
2002	13.2
2003	16.5
2004	19.0
2005	28.6
2006	34.8
2007	45.4
2008	56.6
2009	52.1
2010	83.8
2011	150.8

2-13　海洋工程建筑业增加值
Added Value of Marine Engineering Architecture

单位：亿元　　(100 million yuan)

年　份 Year	增加值 Added Value
2001	109.2
2002	145.4
2003	192.6
2004	231.8
2005	257.2
2006	423.7
2007	499.7
2008	347.8
2009	672.3
2010	874.2
2011	1 086.8

2-14 海洋电力业增加值
Added Value of Marine Electric Power Industry

单位：亿元 (100 million yuan)

年 份 Year	增加值 Added Value
2001	1.8
2002	2.2
2003	2.8
2004	3.1
2005	3.5
2006	4.4
2007	5.1
2008	11.3
2009	20.8
2010	38.1
2011	59.2

2-15 海水利用业增加值
Added Value of Seawater Utilization Industry

单位：亿元 (100 million yuan)

年 份 Year	增加值 Added Value
2001	1.1
2002	1.3
2003	1.7
2004	2.4
2005	3.0
2006	5.2
2007	6.2
2008	7.4
2009	7.8
2010	8.9
2011	10.4

2-16 海洋交通运输业增加值
Added Value of Marine Communications and Transportation Industry

单位：亿元　　(100 million yuan)

年　份 Year	增加值 Added Value
2001	1 316.4
2002	1 507.4
2003	1 752.5
2004	2 030.7
2005	2 373.3
2006	2 531.4
2007	3 035.6
2008	3 499.3
2009	3 146.6
2010	3 785.8
2011	4 217.5

2-17 滨海旅游业增加值
Added Value of Coastal Tourism

单位：亿元　　(100 million yuan)

年　份 Year	增加值 Added Value
2001	1 072.0
2002	1 523.7
2003	1 105.8
2004	1 522.0
2005	2 010.6
2006	2 619.6
2007	3 225.8
2008	3 766.4
2009	4 352.3
2010	5 303.1
2011	6 239.9

2-18 沿海地区海洋生产总值
Gross Ocean Product by Coastal Regions

地 区 Region	海洋生产总值（亿元） Gross Ocean Product (100 million yuan)	第一产业 Primary Industry	第二产业 Secondary Industry	第三产业 Tertiary Industry	海洋生产总值占沿海地区生产总值比重（%） Proportion of the Gross Ocean Product in the Gross Regional Product（%）
合 计 Total	**45 496.0**	**2 381.9**	**21 685.6**	**21 428.5**	**15.7**
天 津 Tianjin	3 519.3	7.2	2 410.4	1 101.7	31.1
河 北 Hebei	1 451.4	61.1	813.7	576.5	5.9
辽 宁 Liaoning	3 345.5	437.1	1 445.7	1 462.7	15.1
上 海 Shanghai	5 618.5	3.8	2 196.8	3 417.9	29.3
江 苏 Jiangsu	4 253.1	135.6	2 297.0	1 820.6	8.7
浙 江 Zhejiang	4 536.8	350.4	2 022.2	2 164.2	14.0
福 建 Fujian	4 284.0	361.4	1 866.0	2 056.6	24.4
山 东 Shandong	8 029.0	540.9	3 961.9	3 526.3	17.7
广 东 Guangdong	9 191.1	225.8	4 311.4	4 654.0	17.3
广 西 Guangxi	613.8	126.8	230.6	256.4	5.2
海 南 Hainan	653.5	131.9	130.0	391.6	25.9

2-19 沿海地区海洋生产总值构成
Composition of Gross Ocean Product by Coastal Regions

单位：%　　　　(%)

地　区 Region	海洋生产总值 Gross Ocean Product	第一产业 Primary Industry	第二产业 Secondary Industry	第三产业 Tertiary Industry
合　计 Total	**100.0**	**5.2**	**47.7**	**47.1**
天　津 Tianjin	100.0	0.2	68.5	31.3
河　北 Hebei	100.0	4.2	56.1	39.7
辽　宁 Liaoning	100.0	13.1	43.2	43.7
上　海 Shanghai	100.0	0.1	39.1	60.8
江　苏 Jiangsu	100.0	3.2	54.0	42.8
浙　江 Zhejiang	100.0	7.7	44.6	47.7
福　建 Fujian	100.0	8.4	43.6	48.0
山　东 Shandong	100.0	6.7	49.3	43.9
广　东 Guangdong	100.0	2.5	46.9	50.6
广　西 Guangxi	100.0	20.7	37.6	41.8
海　南 Hainan	100.0	20.2	19.9	59.9

2-20　沿海地区海洋及相关产业增加值
Added Values of Marine and Related Industries by Coastal Regions

单位：亿元　　(100 million yuan)

地　区 Region	合　计 Total	海洋产业 Marine Industry	主要海洋产业 Major Marine Industry	海洋科研教育管理服务业 Industries of Marine Scientific Research, Education, Management and Service	海洋相关产业 Ocean-related Industries
合　计 Total	**45 496.0**	**26 422.0**	**18 865.2**	**7 556.8**	**19 074.1**
天　津 Tianjin	3 519.3	2 050.1	1 878.6	171.5	1 469.2
河　北 Hebei	1 451.4	762.2	684.5	77.7	689.2
辽　宁 Liaoning	3 345.5	2 050.0	1 642.8	407.2	1 295.6
上　海 Shanghai	5 618.5	3 272.2	2 104.9	1 167.3	2 346.3
江　苏 Jiangsu	4 253.1	2 378.3	1 787.0	591.4	1 874.8
浙　江 Zhejiang	4 536.8	2 560.0	1 822.7	737.3	1 976.8
福　建 Fujian	4 284.0	2 267.2	1 638.0	629.2	2 016.8
山　东 Shandong	8 029.0	4 616.2	3 393.7	1 222.5	3 412.8
广　东 Guangdong	9 191.1	5 612.2	3 273.8	2 338.4	3 578.9
广　西 Guangxi	613.8	386.7	318.3	68.4	227.1
海　南 Hainan	653.5	466.9	320.9	145.9	186.7

2-21 沿海地区海洋及相关产业增加值构成
Composition of the Added Value of Marine and Related Industries by Coastal Regions

单位：% (%)

地 区 Region	合 计 Total	海洋产业 Marine Industry	主要海洋产业 Major Marine Industry	海洋科研教育管理服务业 Industries of Marine Scientific Research, Education, Management and Service	海洋相关产业 Ocean-related Industries
合 计 Total	**100.0**	**58.1**	**41.5**	**16.6**	**41.9**
天 津 Tianjin	100.0	58.3	53.4	4.9	41.7
河 北 Hebei	100.0	52.5	47.2	5.4	47.5
辽 宁 Liaoning	100.0	61.3	49.1	12.2	38.7
上 海 Shanghai	100.0	58.2	37.5	20.8	41.8
江 苏 Jiangsu	100.0	55.9	42.0	13.9	44.1
浙 江 Zhejiang	100.0	56.4	40.2	16.3	43.6
福 建 Fujian	100.0	52.9	38.2	14.7	47.1
山 东 Shandong	100.0	57.5	42.3	15.2	42.5
广 东 Guangdong	100.0	61.1	35.6	25.4	38.9
广 西 Guangxi	100.0	63.0	51.9	11.1	37.0
海 南 Hainan	100.0	71.4	49.1	22.3	28.6

主要统计指标解释

1. 海洋经济 是开发、利用和保护海洋的各类产业活动以及与之相关联活动的总和。

2. 海洋生产总值 是海洋经济生产总值的简称，指按市场价格计算的沿海地区常住单位在一定时期内海洋经济活动的最终成果，是海洋产业和海洋相关产业增加值之和。

3. 海洋产业 是开发、利用和保护海洋所进行的生产和服务活动，包括海洋渔业、海洋油气业、海洋矿业、海洋盐业、海洋化工业、海洋生物医药业、海洋电力业、海水利用业、海洋船舶工业、海洋工程建筑业、海洋交通运输业、滨海旅游业等主要海洋产业以及海洋科研教育管理服务业。

4. 海洋科研教育管理服务业 是开发、利用和保护海洋过程中所进行的科研、教育、管理及服务等活动，包括海洋信息服务业、海洋环境监测预报服务、海洋保险与社会保障业、海洋科学研究、海洋技术服务业、海洋地质勘查业、海洋环境保护业、海洋教育、海洋管理、海洋社会团体与国际组织等。

5. 海洋相关产业 是指以各种投入产出为联系纽带，与主要海洋产业构成技术经济联系的上下游产业，涉及海洋农林业、海洋设备制造业、涉海产品及材料制造业、涉海建筑与安装业、海洋批发与零售业、涉海服务业等。

6. 海洋三次产业 我国的海洋三次产业划分如下：

海洋第一产业：是指海洋渔业中的海洋水产品、海洋渔业服务业，以及海洋相关产业中属于第一产业范畴的部门。

海洋第二产业：是指海洋渔业中海洋水产品加工、海洋油气业、海洋矿业、海洋盐业、海洋化工业、海洋生物医药业、海洋电力业、海水利用业、海洋船舶工业、海洋工程建筑业，以及海洋相关产业中属于第二产业范畴的部门。

海洋第三产业：是指除海洋第一、二产业以外的其他行业。第三产业包括：海洋交通运输业、滨海旅游业、海洋科研教育管理服务业，以及海洋相关产业中属于第三产业范畴的部门。

7. 海洋渔业 包括海水养殖、海洋捕捞、海洋渔业服务业和海洋水产品加工等活动。

8. 海洋油气业 是指在海洋中勘探、开采、输送、加工原油和天然气的生产活动。

9. 海洋矿业 包括海滨砂矿、海滨土砂石、海滨地热与煤矿及深海矿物等的采选活动。

10. 海洋盐业 是指利用海水生产以氯化钠为主要成分的盐产品的活动，包括采盐和盐加工。

11. 海洋船舶工业 是指以金属或非金属为主要材料，制造海洋船舶、海上固定及浮动装置的活动，以及对海洋船舶的修理及拆卸活动。

12. 海洋化工业 包括海盐化工、海水化工、海藻化工及海洋石油化工的化工产品生产活动。

13. 海洋生物医药业 是指以海洋生物为原料或提取有效成分，进行海洋药品与海洋保健品的生产加工及制造活动。

14. 海洋工程建筑业 是指在海上、海底和海岸所进行的用于海洋生产、交通、娱乐、防护等用途的建筑工程施工及其准备活动；包括海港建筑、滨海电站建筑、海岸堤坝建筑、海洋隧道桥梁建筑、海上油气田陆地终端及处理设施建造、海底线路管道和设备安装，不包括各部门、各地区的房屋建筑及房屋装修工程。

15. 海洋电力业 是指在沿海地区利用海洋能、海洋风能进行的电力生产活动。不包括沿海地区的火力发电和核力发电。

16. 海水利用业 是指对海水的直接利用和海水淡化活动，包括利用海水进行淡水生产和将海水应用于工业冷却用水和城市生活用水、消防用水等活动，不包括海水化学资源综合利用活动。

17. 海洋交通运输业 是指以船舶为主要工具从事海洋运输以及为海洋运输提供服务的活动，包括远洋旅客运输、沿海旅客运输、远洋货物运输、沿海货物运输、水上运输辅助活动、管道运输业、装卸搬运及其他运输服务活动。

18. 滨海旅游业 是指以海岸带、海岛及海洋各种自然景观、人文景观为依托的旅游经营、服务活动, 主要包括：海洋观光游览、休闲娱乐、度假住宿、体育运动等活动。

Explanatory Notes on Main Statistical Indicators

1. Marine Economy is the summation of various types of industrial activities for developing, utilizing and protecting the ocean as well as the activities associated with there.

2. Gross Ocean Product is the short form of the gross output value of ocean economy, referring to the final result of marine economic activities of the permanent units in the coastal region within a given period calculated at the market price, and the sum total of the added values of the marine industries as the ocean-related industries.

3. Marine industry refers to the production as service activities for developing, utilizing and protecting the ocean, including major marine industries such as offshore oil and gas industry, marine mining industry, marine salt industry, marine chemical industry, marine biomedicine industry, marine electric power industry, seawater utilization industry, marine shipbuilding industry, marine engineering construction industry, marine communications and transportation industry, coastal tourism etc. as well as marine scientific research, education, management and service.

4. Marine Scientific Research, Education, Management and Service refer to the activities of scientific research, education, management and service carried out in the process of developing, utilizing and protecting the ocean, including marine information service industry, marine environment monitoring and forecasting service, marine insurance and social security industry, marine scientific research, marine technological service industry, ocean geological prospecting industry, marine environmental protection industry, marine education, marine management, marine social organization and international organizations etc.

5. Ocean-Related Industry refers to the lower and upper reaches enterprises that form a technical and economic link with the major marine industries, with various inputs and outputs as ties, involving marine agriculture and forestry, marine equipment manufacturing, ocean-related building and installation industry, marine wholesale and retail industry, ocean-related service industry etc.

6. Marine Three Industries Chinese marine three industries are divided as follows:

Marine primary industry: refers to the marine aquatic products, marine fishery service industry in the marine fishery as well as the sectors belonging to the primary industry category in the ocean-related industries.

Marine secondary industry: refers to the marine aquatic products processing industry in the marine fishery, offshore oil as gas industry, marine mining industry, marine salt industry, marine chemical industry, marine biomedicine industry, marine electric power industry, seawater utilization industry, marine shipbuilding industry, marine engineering construction industry, as well as the sectors belonging to the category of secondary industry in the ocean-related industries.

Marine tertiary Industry: refers to the industries other than the marine primary and secondary industries, including marine communications and transportation industry, coastal tourism, marine scientific research, education, management and service industry as well as the sectors belonging to the category of tertiary industry in the ocean-related industries.

7. Marine Fishery includes mariculture, marine fishing, marine fishery service industry and marine aquatic products processing, etc.

8. Offshore Oil and Gas Industry refers to the production activities of exploring, exploiting, transporting and processing crude oil and natural gas in the ocean.

9. Ocean Mining Industry includes the activities of extracting and dressing beach placers, beach soil and sand, submarine geothermal energy, and coal mining and deep-sea mining, etc.

10. Marine Salt Industry refers to the activity of producing the salt products with the sodium chloride as the main component by utilizing seawater, including salt extracting and processing.

11. Shipbuilding Industry refers to the activity of building ocean vessels, offshore fixed and floating equipment with metals or non-metals as main materials as well as repairing and dismantling ocean vessels.

12. Marine Chemical Industry includes the production activities of chemical products of sea salt, seawater, sea algal and marine petroleum chemical industries.

13. Marine Biomedicine Industry refers to the production, processing and manufacturing activities of marine medicines and marine health care products by using marine organisms as raw materials or extracting useful components therefrom.

14. Marine Engineering Building Industry refers to the architectural projects construction and its preparations in the sea, at the sea bottom and seacoast for such uses as marine production, transportation, recreation, protection, etc., including constructions of seaports, coastal power stations, coastal dykes, marine tunnels and bridges, land terminals of offshore oil and gas fields as well as building of processing facilities, and installation of submarine pipelines and equipment, but not the projects of house building and renovation.

15. Marine Electric Power Industry refers to the activities of generating electric power in the coastal region by making use of ocean energies and ocean wind energy. It does not include the thermal and nuclear power generation in the coastal area.

16. Seawater Utilization Industry refers to the activities of the direct use of sea water and the seawater desalination, including those of carrying out the production of desalination and applying the seawater as water for industrial cooling, urban domestic water, water for fire fighting etc., but not the activity of the multipurpose use of seawater chemical resources.

17. Marine Communications and Transportation Industry refers to the activities of carrying out and serving the sea transportations with vessels as main vehicles, including ocean-going passagers transportation, coastal passagers transportation, ocean-going cargo transportation, coastal cargo transportation, auxiliary activities of water transportation, pipeline transportation, loading, unloading and transport as well as other transportation service activities.

18. Coastal Tourism refers to the tourist business and service activities with the backing of coastal zone, sea islands as well as a variety of natural and human landscapes of the ocean, mainly including marine sightseeing, living a life of leisure and recreation, going on vocation and getting accommodation, sports, etc.

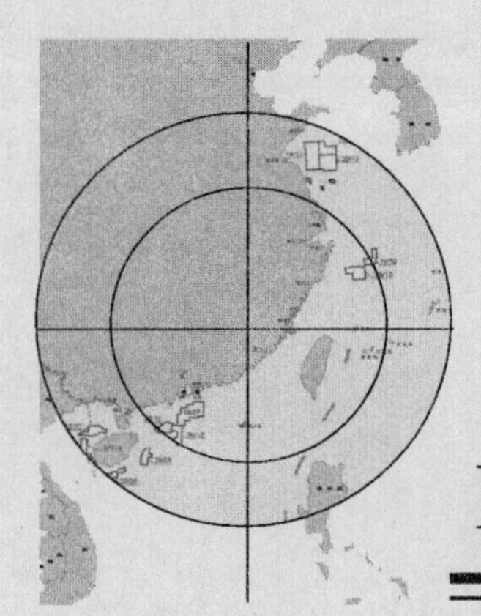

3 主要海洋产业活动

Major Marine Industrial Activities

3-1 全国海洋捕捞养殖产量
National Marine Catches and Mariculture Production

单位：吨 (t)

项　目　Item	2009	2010	2011
海水水产品产量 Total Seawater Aquatic Products	**28 805 399**	**27 975 312** ③	**29 080 487** ③
海洋捕捞产量 Marine Catches	**13 440 772** ①	**12 035 946** ②	**12 419 386** ②
按品种分　By Species			
鱼类　Fish	9 148 657	8 255 051	8 639 947
甲壳类　Crustacea	2 177 651	2 043 314	2 091 282
贝类　Shellfish	681 832	622 104	584 078
藻类　Algae	28 542	24 636	27 362
头足类　Cephalopoda	820 410	658 309	695 251
其他　Others	438 366	432 532	381 466
海水养殖产量 Mariculture Production	**15 364 627**	**14 823 008**	**15 513 292**
按品种分　By Species			
鱼类　Fish	938 601	808 171	964 189
甲壳类　Crustacea	1 102 028	1 061 096	1 127 189
贝类　Shellfish	11 398 288	11 082 321	1 153 626
藻类　Algae	1 549 394	1 541 322	1 601 764
其他　Others	376 316	330 098	276 524

注：①海洋捕捞产量分项中未包括辽宁远洋捕捞产量。②海洋捕捞产量未包括远洋捕捞产量。
③海水水产品产量为海洋捕捞产量、远洋捕捞产量及海水养殖产量之和。

Note: ① Data for Liaoning Deep-Sea fishing production is not included in the marine catches.
② Data for Deep-Sea fishing production is not included in the marine catches.
③ The output of seawater aquatic products is the sum of the productions of marine fishing, deep-sea fishing and mariculture.

3-2 沿海地区海洋捕捞养殖产量
Marine Catches and Mariculture Production by Coastal Regions

单位：吨 (t)

地 区 Region	海洋捕捞产量 Marine Catches	远洋捕捞产量 Deep-Sea Fishing Production	海水养殖产量 Mariculture Production
全国总计 National Total	**12 419 386**	**1 147 809**	**15 513 292**
天 津 Tianjin	17 051	7 986	13 305
河 北 Hebei	251 761		311 520
辽 宁 Liaoning	1 061 607	161 167	2 435 184
上 海 Shanghai	21 457	100 129	
江 苏 Jiangsu	568 108	10 324	842 408
浙 江 Zhejiang	3 030 202	234 703	844 941
福 建 Fujian	1 916 560	183 570	3 161 489
山 东 Shandong	2 384 444	127 993	4 134 775
广 东 Guangdong	1 452 615	73 896	2 655 746
广 西 Guangxi	665 281	4 140	923 804
海 南 Hainan	1 050 300		190 120

3-3 沿海地区海洋原油产量
Output of Offshore Crude Oil by Coastal Regions

单位：万吨　(10 000 t)

地 区 Region	2009	2010	2011
合 计 Total	**3 698.19**	**4 709.98**	**4 451.97**
天 津 Tianjin	1 874.01	2 916.46	2 770.20
河 北 Hebei	200.30	221.19	229.05
辽 宁 Liaoning	15.00	13.01	10.75
上 海 Shanghai	9.68	8.80	16.10
山 东 Shandong	240.01	246.27	257.15
广 东 Guangdong	1 359.19	1 304.25	1 168.72

3-4 沿海地区海洋天然气产量
Output of Offshore Natural Gas by Coastal Regions

单位：万立方米　(10 000 m^3)

地 区 Region	2009	2010	2011
合 计 Total	**859 173**	**1 108 905**	**1 214 519**
天 津 Tianjin	143 002	186 089	213 719
河 北 Hebei	37 043	40 753	50 987
辽 宁 Liaoning	4 575	3 069	2 370
上 海 Shanghai	58 767	49 742	71 789
山 东 Shandong	16 157	12 874	12 280
广 东 Guangdong	599 629	816 378	863 374

3-5 海洋原油出口量及创汇额 Export Volume and Foreign-Exchange Earnings of Offshore Crude Oil by Coastal Regions

单位：万吨，万美元 (10 000 t, 10 000 US$)

地 区 Region	2009		2010		2011	
	出口量 Export Volume	创汇额 Foreign-Exchange Earnings	出口量 Export Volume	创汇额 Foreign-Exchange Earnings	出口量 Export Volume	创汇额 Foreign-Exchange Earnings
合 计 Total	**134.90**	**51 901**	**73.86**	**37 530**	**39.99**	**29 766**
天 津 Tianjin	66.15	24 679	48.28	23 640	25.25	19 009
广 东 Guangdong	68.75	27 222	25.58	13 890	14.74	10 757

3-6 海洋原油产量、出口量占全国原油产量、出口量比重 Proportion of Offshore Crude Oil Production and Export Volume in the National Total

年 份 Year	海洋原油产量占全国原油产量比重（%） Proportion of Offshore Crude Oil Production in the National Total (%)	海洋原油出口量占全国原油出口量比重（%） Proportion of Offshore Crude Oil Export Volume in the National Total (%)
2001	13.07	46.26
2002	14.40	54.95
2003	15.01	60.77
2004	16.16	83.37
2005	17.51	84.18
2006	17.54	89.53
2007	17.06	71.53
2008	17.96	76.46
2009	19.52	26.61
2010	23.27	24.38
2011	21.94	15.87

3-7 沿海地区海洋矿业产量
Output of Marine Mining Industry by Coastal Regions

单位：吨 (t)

地区 Region	产量 Output		
	2009	2010	2011
合　计 Total	**55 906 609**	**34 225 357**	**42 310 427**
浙　江 Zhejiang	47 554 400	25 013 500	27 498 400
福　建 Fujian	2 082 500	2 287 100	2 598 700
山　东 Shandong	3 423 889	4 257 157	9 453 747
广　西 Guangxi	564 820	250 000	260 000
海　南 Hainan	2 281 000	2 417 600	2 499 580

3-8 沿海地区海盐产量
Productive Activities of Salt-Making Industry by Coastal Regions

单位：万吨 (10 000 t)

地 区 Region	海盐产量 Output of Sea Salt		
	2009	2010	2011
合 计 Total	**3 500.45**	**3 286.63**	**3 322.42**
天 津 Tianjin	227.56	204.40	181.00
河 北 Hebei	421.65	429.41	402.25
辽 宁 Liaoning	222.33	146.05	133.62
江 苏 Jiangsu	106.61	149.91	94.25
浙 江 Zhejiang	17.05	10.59	13.99
福 建 Fujian	53.14	29.99	48.50
山 东 Shandong	2 412.72	2 273.05	2 418.63
广 东 Guangdong	11.57	14.16	15.80
广 西 Guangxi	15.34	14.29	3.63
海 南 Hainan	12.48	14.78	10.75

3-9 沿海地区海洋化工产品产量
Output of Marine Chemical Products by Coastal Regions

单位：吨 (t)

地 区 Region	产品产量[①] Output		
	2009	2010	2011
合 计 Total	**12 915 773**	**11 499 913**	**9 202 143**
天 津 Tianjin	1 640 000	1 666 288	1 543 708
河 北 Hebei	820 340	73 960	73 960
辽 宁 Liaoning	446 095[②]	652 876	892 629
江 苏 Jiangsu	1 284 696	1 307 863	73 720
浙 江 Zhejiang	373 133[②]	494 421	619 852
福 建 Fujian	269 840	378 987	752 636
山 东 Shandong	7 240 769	6 349 018	4 580 238
广 东 Guangdong	840 900[②]	576 500[②]	665 400[②]

注：①数据为沿海地区部分海洋化工企业产品汇总数据；②为中国盐业总公司数据。

Note: ① The data collected from the products of part of the chemical enterprises in the coastal region.
② The data from the China Salt Industry Corporation.

3-10 沿海地区海洋生物医药产品产量*
Production of Marine Biomedicine Industry by Coastal Regions

产品名称 Name	计量单位 Unit	产品产量 Output
海参肽营养素胶囊 Sea Cucumber Peptide Nutrient Capsule	箱 box	53 000.00
藻日康 Zaorikang	盒 case	58 500.00
巨藻盖 Juzaogai	盒 case	96 700.00
螺旋藻粉 Spirulina Powder	吨 t	916.00
螺旋藻胶囊 Spirulina Capsule	瓶 bottle	6 000.00
螺旋藻胶囊 Spirulina Capsule	万盒 10 000 cases	5 779.10
藻酸双酯钠片 Alginic Acid Diadipose Sodium Pill	万片 10 000 pills	237.00
藻酸双酯钠（针剂） Alginic Acid Diadipose Sodium Drops	万支 10 000 bottles	34.00
鲨鱼肝油胶丸 Shark Liver Oil Pill	万粒 10 000 pellets	7 463.00
金枪鱼油胶丸 Tuna Oil Pill	万粒 10 000 pellets	2 027.00
卵磷脂胶丸 Lecithin Pill	万粒 10 000 pellets	821.00
鳕鱼肝油胶丸 Ling Liver Oil Pill	万粒 10 000 pellets	335.00
海藻植物胶囊 Algae Plant Capsule	吨 t	340.00
蚝贝钙片 Oyster Shell Calcium Tablets	万片 10 000 pills	8 945.98
鱼油软胶囊 Fish Oil Soft Capsule	万粒 10 000 pellets	1 208.00
鱼油胶囊 Fish Oil Capsule	吨 t	11 076.00
鱼肝油乳 Cod-Liver Oil Milk	万瓶 10 000 bottles	582.77
鱼肝油系列产品 Cod-Liver Oil series products	箱 box	2 000.00
鱼肝油系列产品 Cod-Liver Oil series products	吨 t	1 978.00
鲨鱼硫酸软骨素 Shark Sulphate Chondroitin	吨 t	436.67
海蛇痹宁胶囊 Sea Snake Anti-rheumatic Capsule	万粒 10 000 pellets	560.23
维生素AD胶丸 Vitamin AD Capsule	万粒 10 000 pellets	90 826.00

3-10 续表1 continued

产品名称 Name	计量单位 Unit	产品产量 Output
维生素AD滴剂 Vitamin AD Drops	万支 10 000 bottles	8 637.62
维生素E胶丸 Vitamin E Capsule	万粒 10 000 pellets	37 033.46
鲎试剂 King Crab Tablet	万支 10 000 bottles	17.00
海珠喘息定片 Haizhu Methoxyphenamine Pill	万片 10 000 pills	19 414.00
珍珠粉末 Pearl Powder	万支 10 000 bottles	464.07
金牡感冒片 Jinmu Cold Cure Pill	万片 10 000 pills	1 018.54
多烯酸乙酯 Ethyl Polyenoic Acid	吨 t	9.00
多烯酸乙酯胶囊 Ethyl Polyenoic Acid Capsule	万粒 10 000 pellets	445.83
鲨力胶囊 Shali Capsule	瓶 bottle	10 000.00
氨糖美辛肠溶片 Glucosamine Indometacin Enteric-coated Tablets	万片 10 000 pills	480.48
去甲斑蝥素片 Demethylcantharidin Tablets	万片 10 000 pills	109.78
珍珠胶囊口服液 Pearl Capsule Oral Liquid	万支 10 000 bottles	400.00
甘糖酯 Mannose Easter	万片 10 000 pills	320.00
珠珀惊风散 Zhubo Convulsions Powder	万瓶 10 000 bottles	421.40
康体通胶囊 Kangtitong Capsule	万粒 10 000 pellets	5 000.00
石决明 Shell of Abalone	公斤 kg	93.00
海螵蛸 Cuttlebone	公斤 kg	529.50
珍珠母 Mother-of-pearl	公斤 kg	186.00
昆布 Kelp	公斤 kg	150.00
龟板 Tortoise Shell	公斤 kg	80.00
鳖甲 Turtle Shell	公斤 kg	250.00
鱼蛋白粉 Fish Protein Powder	吨 t	45.20
甲壳素 Crustaceoxin	吨 t	2 620.00
脑元神软胶囊 Naoyuanshen Soft Capsule	万粒 10 000 pellets	22.30
海麟舒肝胶囊 Hailin Liver-Soothing Capsule	万粒 10 000 pellets	26.00
海威口服液 Haiwei Oral Liquid	吨 t	402.00

3-10 续表2 continued

产品名称 Name	计量单位 Unit	产品产量 Output
伊可新 Vitamin A and D Drops	吨 t	164.00
苯海因 Benzene Glycolylurea	吨 t	4 506.00
乙酰胺基噻二唑 Acetazolamide	吨 t	290.00
卡拉胶 Carrageenan	吨 t	3 153.00
角鲨烯胶囊 Houndfish Alkene Capsule	万粒 10 000 pellets	638.23
D酯 D ester	吨 t	5 390.00
麝珠明目滴眼液 Shezhu Eye-brightening Drops	万瓶 10 000 bottles	166.00
碘酊 Iodo-tincture	万瓶 10 000 bottles	7.64
碘甘油 Iodoglycerol	万瓶 10 000 bottles	3.66
鱼胶原蛋白肽 Fish Collagen Peptide	万粒 10 000 pellets	123.21
八宝惊风散 Babao Infantile Convulsions Powder	万盒 10 000 cases	135.51
海藻酸钠 Sodium Algrnic Acid	吨 t	182 364.00
碘 Iodine	吨 t	2 210.00
海狗油软胶囊 Fur Seal Oil Soft Capsule	瓶 bottles	92 779.00
金枪鱼软胶囊 Tuna Soft Capsule	万粒 10 000 pellets	213.00
甘露醇 Mannitol	吨 t	278.00
二巯基1，3，4噻二唑 Disulfenyl 1, 3, 4 thiadiazole	吨 t	6.00
肤疹宁软膏 Rash Cream	万支 10 000 bottles	3.44
保健品鱼油 Health Care Fish Oil	吨 t	2 167.00
角鲨烯原料 Squalene Raw Material	吨 t	8.00
南海岸鳗钙系列产品 South Coast Eel-calcium	万盒 10 000 cases	129.88
伤科接骨片 Traumatologic Osteopathic Tablets	吨 t	372.44
鲨鱼软骨罐头 Shark Cartilage Can	罐 can	32 000.00

注：数据为沿海地区部分海洋生物医药产品汇总数据。

Note: The data collected from the products of part of the marine biomedicine enterprises in the coastal region.

3-11 沿海地区海洋修造船完工量
Production of the Marine Shipbuilding Industry by Coastal Regions

部门和地区 Sector and Region	修船完工量（艘） Ships Repaired (number)	造船完工量 Ships Built	
		艘 number	万综合吨 Comprehensive Tonnages (10 000 t)
总　计　Total	**10 501**	**3 412**	**7 323.19**
其　中: Including:			
中船工业集团公司 CSSC	547	176	1 744.73
中船重工集团公司 CSIC	494	96	1 080.15
按地区分: By Regions:			
天　津 Tianjin	185	23	33.26
河　北 Hebei	580	13	90.90
辽　宁 Liaoning	220	145	1 100.00
上　海 Shanghai	1 127	108	1 340.88
江　苏 Jiangsu	531	1 238	2 703.00
浙　江 Zhejiang	3 918	834	1 144.61
福　建 Fujian	1 794	561	148.66
山　东 Shandong	1 600	339	357.10
广　东 Guangdong	444	68	403.85
广　西 Guangxi	67	12	0.19
海　南 Hainan	35	71	0.74

3-12 沿海地区海洋货物运输量和周转量
Maritime Volume of Goods Transported and Turnover by Coastal Regions

单位：万吨，亿吨千米 (10 000 t, 100 million t-km)

地 区 Region	货运量 Volume of Goods Transported	沿 海 Coastal	远 洋 Oceangoing	货物周转量 Volume of Goods Turnover	沿 海 Coastal	远 洋 Oceangoing
全国总计 National Total	**205 808**	**142 991**	**62 817**	**67 551**	**18 635**	**48 915**
天 津 Tianjin	12 717	3 299	9 418	9 550	581	8 969
河 北 Hebei	2 672	2 672	0	495	495	0
辽 宁 Liaoning	11 615	5 347	6 268	6 529	799	5 731
上 海 Shanghai	46 929	30 885	16 044	19 953	4 299	15 654
江 苏 Jiangsu	16 229	10 987	5 242	4 489	1 081	3 407
浙 江 Zhejiang	43 734	41 843	1 891	6 460	5 008	1 453
福 建 Fujian	16 690	14 914	1 776	2 546	2 045	501
山 东 Shandong	11 697	6 812	4 885	4 246	676	3 570
广 东 Guangdong	24 021	14 889	9 132	4 070	2 178	1 893
广 西 Guangxi	3 787	3 326	461	615	597	18
海 南 Hainan	9 126	8 017	1 109	1 238	878	360
其 他* Others	6 591	0	6 591	7 359	0	7 359

注：其他数据为中国远洋运输（集团）总公司完成的远洋运输量（表3-15同）。

Note: Other data are those of the freight volume of ocean transportation accomplished by the China National Ocean Shipping Corporation (Same as Table 3-15).

3-13 沿海地区海洋旅客运输量和周转量
Maritime Volume of Passenger Traffic and Turnover by Coastal Regions

单位：万人，亿人千米 (10 000 persons, 100 million person-km)

地 区 Region	客运量 Passenger Traffic	沿 海 Coastal	远 洋 Oceangoing	旅客周转量 Passenger Turnover Volume	沿 海 Coastal	远 洋 Oceangoing
全国总计 National Total	**10 473**	**9 544**	**929**	**41.15**	**30.58**	**10.57**
天 津 Tianjin	1	0	1	0.17	0.00	0.17
辽 宁 Liaoning	549	536	13	7.04	6.42	0.62
上 海 Shanghai	358	358	0	1.02	1.02	0.00
江 苏 Jiangsu	25	25	0	0.86	0.86	0.00
浙 江 Zhejiang	2 603	2 603	0	5.52	5.52	0.00
福 建 Fujian	1 349	1 271	78	2.00	1.48	0.51
山 东 Shandong	1 952	1 859	93	11.61	7.84	3.78
广 东 Guangdong	2 273	1 530	743	9.02	3.57	5.46
广 西 Guangxi	153	152	1	0.87	0.85	0.03
海 南 Hainan	1 210	1 210	0	3.02	3.02	0.00

3-14 沿海港口客货吞吐量
Passengers Leaving and Arriving and Cargo Handled at Coastal Seaports

单位：万吨，万人次 (10 000 t, 10 000 person-times)

地区 Region	货物吞吐量 Cargo Handled	#外贸 Foreign Trade	旅客吞吐量 Passenger Leaving & Arriving	#离港 Leaving
合计 Total	**636 024**	**254 402**	**7 999**	**4 048**
天津 Tianjin	45 338	22 162	25	12
河北 Hebei	71 300	16 354	6	3
辽宁 Liaoning	78 344	17 369	704	349
上海 Shanghai	62 432	33 778	154	78
江苏 Jiangsu	17 752	9 356	14	7
浙江 Zhejiang	86 700	33 613	1 062	526
福建 Fujian	37 279	15 171	1 113	556
山东 Shandong	96 188	53 449	1 237	620
广东 Guangdong	114 455	42 238	2 358	1 226
广西 Guangxi	15 331	8 906	30	15
海南 Hainan	10 905	2 008	1 297	654

3-15 沿海地区水路国际标准集装箱运量
Volume of International Standardized Containers Traffic by Coastal Regions

单位：万TEU，万吨 (10 000 TEU, 10 000 t)

地 区 Region	2009		2010		2011	
	箱 数 Containers	重 量 Weight	箱 数 Containers	重 量 Weight	箱 数 Containers	重 量 Weight
合 计 Total	**2 905**	**33 143**	**3 659**	**40 256**	**4 074**	**49 348**
天 津 Tianjin	12	68	12	177	8	85
河 北 Hebei	1	20	1	24	1	36
辽 宁 Liaoning	39	393	44	444	45	521
上 海 Shanghai	1 332	17 710	1 623	20 561	1 759	22 981
江 苏 Jiangsu	331	2 702	392	3 283	462	3 961
浙 江 Zhejiang	80	1 529	102	1 515	168	2 781
福 建 Fujian	205	3 272	231	3 902	268	4 673
山 东 Shandong	132	1 416	166	2 227	123	2 345
广 东 Guangdong	704	5 063	913	6 812	887	7 342
广 西 Guangxi	49	675	77	992	181	2 367
海 南 Hainan	22	295	100	317	172	2 256

注：国际标准集装箱运量数据包含内河集装箱运量。

Note: The data in this table include the volume of containers transported in the inland rivers.

3-16 沿海港口国际标准集装箱吞吐量
International Standardized Containers Handled at Coastal Seaports

单位：万TEU，万吨 (10 000TEU, 10 000 t)

地　区 Region	2009		2010		2011	
	箱　数 Containers	重　量 Weight	箱　数 Containers	重　量 Weight	箱　数 Containers	重　量 Weight
合　计 Total	**11 020**	**114 415**	**13 145**	**137 076**	**14 632**	**157 969**
天　津 Tianjin	870	8 871	1 009	10 916	1 159	11 929
河　北 Hebei	57	869	62	997	77	1 251
辽　宁 Liaoning	812	11 718	969	15 666	1 200	19 115
上　海 Shanghai	2 500	24 619	2 907	27 992	3 174	31 220
江　苏 Jiangsu	305	2 872	391	3 810	488	4 626
浙　江 Zhejiang	1 118	9 839	1 404	12 276	1 584	15 747
福　建 Fujian	716	9 001	867	10 743	970	12 099
山　东 Shandong	1 312	13 817	1 531	15 189	1 691	17 738
广　东 Guangdong	3 236	31 432	3 868	37 413	4 103	41 315
广　西 Guangxi	35	555	56	926	74	1 165
海　南 Hainan	59	822	82	1 147	112	1 764

3-17 沿海城市国内旅游人数
Domestic Visitors by Coastal Cities

单位：万人次 (10 000 person-times)

城 市	City	2008	2009	2010
合 计	**Total**	**74 699**	**91 807**	**107 019**
天 津	**Tianjin**	7 004	5 537	6 118
河 北	**Hebei**	2 553	3 362	3 968
唐 山	Tangshan	957	1 226	1 532
秦皇岛	Qinhuangdao	1 227	1 638	1 861
沧 州	Cangzhou	369	498	575
辽 宁	**Liaoning**	7 735	9 800	11 655
大 连	Dalian	3 000	3 412	3 777
丹 东	Dandong	1 420	1 897	2 248
锦 州	Jinzhou	870	1 191	1 442
营 口	Yingkou	585	820	1 074
盘 锦	Panjin	980	1 270	1 654
葫芦岛	Huludao	880	1 211	1 460
上 海	**Shanghai**	11 006	12 361	21 463
江 苏	**Jiangsu**	3 146	3 655	4 255
南 通	Nantong	1 275	1 483	1 757
连云港	Lianyungang	1 065	1 210	1 393
盐 城	Yancheng	806	962	1 105
浙 江	**Zhejiang**	19 229	21 948	26 371
杭 州	Hangzhou	4 552	5 094	6 305
宁 波	Ningbo	3 465	3 962	4 624
温 州	Wenzhou	2 547	2 931	3 537
嘉 兴	Jiaxing	2 140	2 492	3 070
绍 兴	Shaoxing	2 435	2 851	3 436
舟 山	Zhoushan	1 495	1 731	2 113
台 州	Taizhou	2 595	2 887	3 286
福 建	**Fujian**		6 938	8 262
福 州	Fuzhou		1 938	2 275
厦 门	Xiamen		1 788	2 178
莆 田	Putian		632	775
泉 州	Quanzhou		1 171	1 351
漳 州	Zhangzhou		835	1 002
宁 德	Ningde		574	681

3-17 续表 continued

城　市 City		2008	2009	2010
山　东	**Shandong**	11 490	13 580	16 055
青　岛	Qingdao	3 390	3 903	4 397
东　营	Dongying	428	515	637
烟　台	Yantai	2 346	2 763	3 272
潍　坊	Weifang	1 869	2 313	2 946
威　海	Weihai	1 586	1 839	2 112
日　照	Rizhao	1 451	1 724	2 031
滨　州	Binzhou	421	523	661
广　东	**Guangdong**	10 170	11 696	13 612
广　州	Guangzhou	2 916	3 286	3 692
深　圳	Shenzhen	1 790	1 944	2 265
珠　海	Zhuhai	829	911	1 055
汕　头	Shantou	604	668	769
江　门	Jiangmen	728	746	872
湛　江	Zhanjiang	165	462	602
茂　名	Maoming	171	201	305
惠　州	Huizhou	675	799	913
汕　尾	Shanwei	219	239	327
阳　江	Yangjiang	240	273	305
东　莞	Dongguan	994	1 191	1 289
中　山	Zhongshan	463	504	540
潮　州	Chaozhou	219	274	317
揭　阳	Jieyang	157	198	361
广　西	**Guangxi**	1 191	1 630	1 958
北　海	Beihai	695	810	938
防城港	Fangchenggang	150	418	550
钦　州	Qinzhou	346	402	469
海　南	**Hainan**	1 175	1 299	1 564
海　口	Haikou	622	662	723
三　亚	Sanya	553	637	841

注：数据来源于《中国区域经济统计年鉴》（2011）。

Note：The data come from *China Statistical Yearbook For Regional Economy (2011)*.

3-18 主要沿海城市国际旅游（外汇）收入
International Tourism (Foreign Exchange) Receipts in Major Coastal Cities

单位：万美元 (10 000 US$)

城 市	City	2009	2010	2011
天 津	Tianjin	118 264	141 951	175 553
秦皇岛	Qinhuangdao	11 909	12 022	13 770
大 连	Dalian	72 748	80 386	80 519
上 海	Shanghai	474 402	634 092	575 118
南 通	Nantong	30 933	36 066	39 916
连云港	Lianyungang	9 173	10 747	12 869
杭 州	Hangzhou	137 995	169 008	195 710
宁 波	Ningbo	48 650	59 066	65 472
温 州	Wenzhou	17 797	21 115	25 602
福 州	Fuzhou	77 400	84 300	102 854
厦 门	Xiamen	90 194	108 552	129 901
泉 州	Quanzhou	64 771	66 737	79 661
漳 州	Zhangzhou	12 685	15 455	18 781
青 岛	Qingdao	55 178	60 103	68 933
烟 台	Yantai	31 081	37 707	46 816
威 海	Weihai	16 083	19 151	21 855
广 州	Guangzhou	362 396	466 127	485 306
深 圳	Shenzhen	276 026	315 896	374 474
珠 海	Zhuhai	102 670	122 339	106 685
汕 头	Shantou	4 910	5 016	5 071
湛 江	Zhanjiang	2 237	2 716	3 630
中 山	Zhongshan	20 434	27 591	24 717
北 海	Beihai	1 721	2 173	2 574
海 口	Haikou	3 089	3 756	3 845
三 亚	Sanya	19 497	24 504	31 259

3-19 主要沿海城市接待入境旅游者人数
Number of Inbound Tourist Arrivals in Major Coastal Cities

单位：人次 (person-time)

城 市 City		2009	2010	2011
天 津	Tianjin	1 410 244	1 660 682	730 615
秦皇岛	Qinhuangdao	224 206	242 337	264 372
大 连	Dalian	1 050 043	1 166 020	1 170 035
上 海	Shanghai	5 333 935	7 337 216	6 686 144
南 通	Nantong	299 866	355 133	404 852
连云港	Lianyungang	100 076	116 663	132 289
杭 州	Hangzhou	2 304 045	2 757 147	3 063 140
宁 波	Ningbo	800 548	951 680	1 073 872
温 州	Wenzhou	329 734	391 587	470 504
福 州	Fuzhou	629 314	698 607	761 665
厦 门	Xiamen	1 281 907	1 551 865	1 799 205
泉 州	Quanzhou	626 721	770 457	846 389
漳 州	Zhangzhou	219 382	247 469	293 728
青 岛	Qingdao	1 000 670	1 080 511	1 156 391
烟 台	Yantai	400 901	472 023	548 533
威 海	Weihai	322 676	372 646	415 114
广 州	Guangzhou	6 894 044	8 147 900	7 786 900
深 圳	Shenzhen	8 963 697	10 206 000	11 045 500
珠 海	Zhuhai	2 978 421	3 251 400	3 208 300
汕 头	Shantou	123 161	133 800	140 700
湛 江	Zhanjiang	50 665	103 200	142 300
中 山	Zhongshan	475 769	480 600	608 000
北 海	Beihai	61 437	73 008	83 073
海 口	Haikou	103 206	132 877	146 808
三 亚	Sanya	317 833	415 939	528 942

3-20 主要沿海城市接待入境旅游者情况
Breakdown of Inbound Tourists in Major Coastal Cities

单位：人次，人天 (person-time, night)

城市 City	外国人 Foreigners		香港同胞 Hong Kong	
	人次数 Arrivals	人天数 Nights	人次数 Arrivals	人天数 Nights
天　津 Tianjin	635 795	7 076 899	44 092	715 888
秦皇岛 Qinhuangdao	248 499	945 509	7 138	23 056
大　连 Dalian	1 038 050	3 514 077	60 305	197 375
上　海 Shanghai	5 549 900	18 880 759	479 536	1 511 649
南　通 Nantong	359 306	1 955 191	17 143	107 631
连云港 Lianyungang	107 583	659 916	8 044	54 219
杭　州 Hangzhou	2 108 263	6 069 395	563 844	1 935 017
宁　波 Ningbo	608 675	1 859 894	170 676	454 521
温　州 Wenzhou	372 442	1 069 890	40 041	95 500
福　州 Fuzhou	428 372	3 341 677	92 089	632 660
厦　门 Xiamen	618 108	3 019 722	166 273	723 534
泉　州 Quanzhou	125 894	760 896	476 818	2 492 163
漳　州 Zhangzhou	67 320	275 367	88 133	325 382
青　岛 Qingdao	807 448	2 120 265	173 594	601 368
烟　台 Yantai	444 674	1 957 041	29 136	69 842
威　海 Weihai	388 471	966 661	3 074	6 848
广　州 Guangzhou	2 762 800	8 001 500	4 026 900	9 532 300
深　圳 Shenzhen	1 712 000	3 682 500	8 818 300	17 547 600
珠　海 Zhuhai	581 600	1 335 800	1 216 300	2 340 600
汕　头 Shantou	87 300	168 400	43 600	78 400
湛　江 Zhanjiang	66 700	166 500	60 300	130 900
中　山 Zhongshan	114 500	464 800	337 900	845 500
北　海 Beihai	41 703	74 410	28 917	45 517
海　口 Haikou	76 708	138 104	28 485	44 847
三　亚 Sanya	421 201	1 489 470	75 266	154 442

3-20 续表 continued

城 市 City	澳门同胞 Macao		台湾同胞 Taiwan Province	
	人次数 Arrivals	人天数 Nights	人次数 Arrivals	人天数 Nights
天 津 Tianjin	2 439	46 484	48 289	711 467
秦皇岛 Qinhuangdao	659	2 155	8 076	26 248
大 连 Dalian	1 737	6 248	69 943	229 496
上 海 Shanghai	24 244	84 840	632 464	2 370 868
南 通 Nantong	1 969	9 403	26 434	170 742
连云港 Lianyungang	3 728	8 999	12 934	108 447
杭 州 Hangzhou	30 742	93 307	360 291	973 685
宁 波 Ningbo	54 717	166 874	239 804	612 936
温 州 Wenzhou	13 621	21 390	44 400	102 935
福 州 Fuzhou	6 437	49 856	234 767	819 311
厦 门 Xiamen	6 234	27 668	1 008 590	2 840 520
泉 州 Quanzhou	46 200	212 320	197 477	768 589
漳 州 Zhangzhou	8 737	25 465	129 538	375 324
青 岛 Qingdao	39 001	123 561	136 348	419 708
烟 台 Yantai	16 096	37 758	58 627	160 288
威 海 Weihai	802	1 826	22 767	65 600
广 州 Guangzhou	462 100	1 033 700	535 100	1 510 400
深 圳 Shenzhen	50 400	93 600	464 800	1 178 900
珠 海 Zhuhai	717 600	1 704 900	692 800	1 457 500
汕 头 Shantou	700	1 500	9 100	17 300
湛 江 Zhanjiang	4 700	10 900	10 600	24 200
中 山 Zhongshan	101 900	258 400	53 700	197 600
北 海 Beihai	3 833	6 292	8 620	15 238
海 口 Haikou	1 137	1 742	40 478	56 120
三 亚 Sanya	4 614	9 383	27 861	54 316

主要统计指标解释

1. 海洋捕捞产量 凡是从海洋里捕捞的天然生长的水产品产量为捕捞产量。

2. 海水养殖产量 凡是从人工投放苗种或天然纳苗并进行人工饲养管理的海水养殖水域中捕捞的水产品产量为海水养殖产量。

3. 远洋捕捞产量 由各远洋渔业企业和各生产单位按我国远洋渔业项目管理办法组织的远洋渔船（队）在非我国管辖海域（外国专属经济区水域或公海）捕捞的水产品产量。中外合资、合作渔船捕捞的水产品只统计按协议应属于中方所有的部分。

4. 原油产量 是按净原油量来计算的，能直接用于销售和生产自用的原油量。目前海洋石油系统原油产量计算方法采用倒算法。

原油产量=销售量+期末库存量-期初库存量+海上平台及陆地终端处理厂自用量。

5. 天然气产量 指进入集输管网的销售量和就地利用的全部气量。

天然气产量=外输（销）量+企业自用量

6. 海洋原油出口量 指销往国外的产品数量。

7. 海洋原油出口创汇额 指产品销往国外的归中方所有的全部外汇收入。以美元或万美元表示。

8. 造船综合吨 等于以计量单位载重吨和满载排水量吨的民用船舶的吨位数之和。

9. 货运量 指经船舶实际运送的货物重量，按到达量统计。

10. 货物周转量 指实际运送的货物与其运送距离的乘积。

11. 集装箱运量 既包含货重，也包含箱重。箱重系指承运租用的空箱重量凡有运费收入的空箱，其重量应统计为运量，按空箱 1 吨为货运量 1 吨计算；若无收入，所承运的空箱一律不作运量统计。

12. 旅客周转量 指实际运送的旅客人数与其运送距离的乘积。

13. 国际旅游外汇收入 入境游客在中国（大陆）境内旅行、游览过程中用于交通、参观游览、住宿、餐饮、购物、娱乐等的全部花费。

14. 接待人次数 指报告期内我国接待游客人数。游客按出游地分为入境游客和国内游客，按出游时间分为旅游者（过夜游客）和一日游游客（不过夜游客）。

15. 接待人天数 指过夜旅游者的停留天数。

16. 外国人 指外国国籍的人，加入外籍的中国血统华人也计入外国人。

17. 港澳台同胞 指居住在我国香港特别行政区、澳门特别行政区和台湾省的中国同胞。

Explanatory Notes on Main Statistical Indicators

1. Marine Catches refers to the output of the naturally growing aquatic products caught from the sea.

2. Mariculture Production refers to the output of aquatic products whose young are artificially released or naturally collected, and raised and managed artificially, and which are caught from the waters of mariculture.

3. Deep-Sea Fishing Production refers to the output of aquatic products caught in the non-Chinese jurisdictional sea areas (foreign EEE or high sea) by the distant fishing vessels (fleet) organized by various distant fishing businesses and production units according to the management measures of the China distant fishing projects. The aquatic products caught by the Chinese-foreign joint ventures' and cooperative fishing vessels are counted only for the part owned by the Chinese side according to the agreement.

4. Output of Crude Oil is calculated on the basis of the net amount of crude oil, i.e., the amount of crude oil that may be directly used for sale and for the production itself.

Output of crude oil = Volume of sales + Reserves at the end of the period - Reserves at the beginning of the period +Amount for self-use on the platforms and in the terminal processing plants on land.

5. Output of Natural Gas refers to the total gas volume of the sales volume entering the oil collecting and transport pipeline network and that used locally.

Output of natural gas = Volume of sales or transport to other areas + Volume used by the enterprise itself

6. Export Volume of Offshore Crude Oil refers to the amount of products for sale abroad.

7. Foreign Exchange Earnings of Offshore Crude Oil refer to the total foreign exchange income from oil (gas) products for sale abroad which is owned by the Chinese side.

8. Comprehensive Tonnages of Shipbuilding refers to the sum of tonnage of civilian vessels with the deadweight capacity and full-load displacement as measured.

9. Freight Traffic refers to the weight of cargoes actually transported by vessels, which is counted according to the volume of arrival.

10. Cargoes Turnover Volume refers to the product of the actually transported cargoes and the transport distance.

11. Freight Volume of Containers includes the weight of both cargoes and container boxes. The weight of container boxes refers to the weight of empty containers rented for transport or having freight income and should be included in the freight volume, one ton of empty boxes equalling to one ton of freight volume. The empty boxes which have no income for transportation are not included in the freight volume.

12. Passenger Turnover Volume refers to the product of the number of passengers actually transported and the shipping distance.

13. International Tourism (Foreign Exchange) Receipts refer to the total expenditure made by inbound visitors within the territory of China (the mainland) in their course of travel on transport, tours and sightseeing, lodging, food and beverage, shopping, entertainment, etc.

14. Number of Person-Times Received refers to the number of visitors received by China in the period reported. Visitors are divided into inbound visitors and domestic visitors by origin of the travel, and tourists (overnight visitors) and same-day visitors by their length of stay.

15. Number of the Days of Stay refers to the number of the days of stay of tourists.

16. Foreigners refer to the people with foreign nationality, including foreign nationals of Chinese descent.

17. Compatriots from Hong Kong, Macao and Taiwan Province refer to the Chinese compatriots living in the Hong Kong Special Administrative Region, the Macao Special Administrative Region and Taiwan Province.

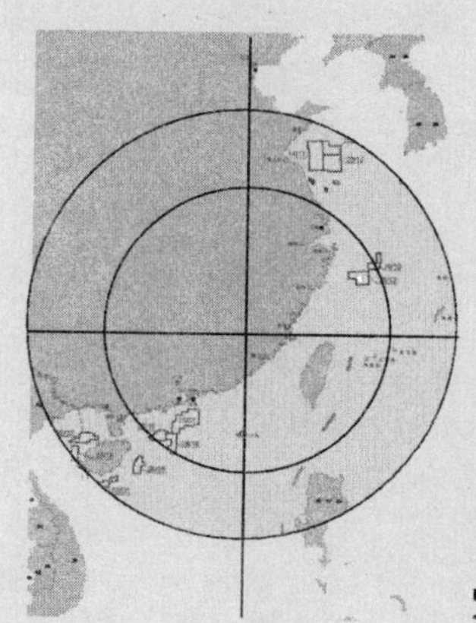

4

主要海洋产业生产能力

Production Capacity of Major Marine Industries

4-1 沿海地区渔港情况
Fishing Ports in the Coastal Regions

单位：个 (number)

地 区	Region	渔港合计 The Total of the Fishing Ports	中心渔港 Central Fishing Port	一级渔港 Grade 1 Fishing Port
全国总计	**National Total**	**126**	**59**	**67**
天 津	Tianjin			
河 北	Hebei	7	3	4
辽 宁	Liaoning	9	3	6
上 海	Shanghai			
江 苏	Jiangsu	10	6	4
浙 江	Zhejiang	17	8	9
福 建	Fujian	17	7	10
山 东	Shandong	17	8	9
广 东	Guangdong	15	7	8
广 西	Guangxi	8	4	4
海 南	Hainan	10	6	4

资料来源：《2012中国渔业统计年鉴》。

Source: *China Fishery Statistical Yearbook 2012*.

4-2 沿海地区海水养殖面积
Mariculture Area by Coastal Regions

单位：公顷 (hm^2)

地 区 Region	海水养殖面积 Mariculture Area
合 计 Total	**2 106 382**
天 津 Tianjin	4 110
河 北 Hebei	134 264
辽 宁 Liaoning	751 387
上 海 Shanghai	
江 苏 Jiangsu	201 073
浙 江 Zhejiang	90 839
福 建 Fujian	142 315
山 东 Shandong	512 126
广 东 Guangdong	203 410
广 西 Guangxi	52 212
海 南 Hainan	14 646

4-3 海洋油田生产井情况
Survey of Offshore Oilfield Production Wells

单位：口 (number)

地 区 Region	合 计 Total	采油井 Oil Wells	采气井 Gas Wells	注水井 Injection Wells	其他井 Others
合 计 Total	**5 156**	**3 900**	**233**	**1 012**	**11**
天 津 Tianjin	2 982	2 257	107	618	
河 北 Hebei	779	598		181	
辽 宁 Liaoning	244	192	6	35	11
上 海 Shanghai	39	13	26		
山 东 Shandong	568	404	8	156	
广 东 Guangdong	544	436	86	22	

4-4 海洋石油勘探情况
Work Volume of Offshore Oil Exploration

地 区 Region	地震测线 Seismic Line		钻井（口） Drilling (Well)	
	二维 （千米） Two Dimensions (km)	三维 （平方千米） Three Dimensions (km^2)	预探井 Wildcat Wells	评价井 Appraisal Wells
合 计 Total	**20 187**	**16 252**	**93**	**70**
天 津 Tianjin	0	6 595	25	29
河 北 Hebei	1 273	0	21	14
辽 宁 Liaoning	0	0	4	1
上 海 Shanghai	6 709	1 107	4	1
山 东 Shandong		430	24	2
广 东 Guangdong	12 205	8 120	15	23
其中：合作 Including: Cooperative	3 284	7 640	9	2

4-5 沿海地区盐田面积和海盐生产能力
Salt Pan Area and Sea Salt Production Capacity by Coastal Regions

地 区 Region	盐田总面积 （公顷） Total Area of Salt Pan (hm^2)		生产面积 （公顷） Production Area (hm^2)		年末海盐生产能力 （万吨） Year-End Capacity of Sea Salt Production (10 000 t)	
	2010	2011	2010	2011	2010	2011
合 计 Total	**473 068**	**455 433**	**332 099**	**359 264**	**3 960.00**	**3 757.25**
天 津 Tianjin	30 512	28 123	29 050	27 461	180.00	166.56
河 北 Hebei	80 359	71 456	65 243	63 930	476.00	461.10
辽 宁 Liaoning	39 790	33 059	33 533	28 415	234.00	207.22
江 苏 Jiangsu	65 634	71 447	19 204	42 365	160.00	110.13
浙 江 Zhejiang	3 355	2 790	2 703	2 303	16.00	14.52
福 建 Fujian	5 607	5 607	4 893	4 893	45.00	53.48
山 东 Shandong	230 122	225 871	166 962	179 518	2 800.00	2 700.00
广 东 Guangdong	10 052	10 052	6 014	6 014	19.00	18.53
广 西 Guangxi	3 949	3 373	1 655	1 557	11.00	7.39
海 南 Hainan	3 688	3 655	2 842	2 808	19.00	18.32

4-6 沿海地区风能发电能力
Wind Power Production by Coastal Regions

单位：万千瓦 (10 000 kW)

地 区 Region	风能年发电能力 Annual Wind Power Generation Capacity
合 计 **Total**	**1 407.77**
天 津 Tianjin	24.35
辽 宁 Liaoning	114.45
河 北 Hebei	20.55
上 海 Shanghai	318.00
江 苏 Jiangsu	195.51
浙 江 Zhejiang	35.22
福 建 Fujian	102.57
山 东 Shandong	441.16
广 东 Guangdong	130.04
广 西 Guangxi	0.25
海 南 Hainan	25.67

4-7　主要潮汐电站分布情况
Distribution of Major Tidal Power Stations

电站名称 Name	运行情况 Status of operation	装机容量（千瓦） Capacity (kW)
江厦潮汐试验电站	1970年开始建造，1980年投入使用，运行至今。	3 900
Jiangxia Experimental Tidal Power Station	It began construction in 1970, was put into use in 1980, and has been in operation so far.	
海山潮汐电站	1972年开始建造，1975年投入使用，运行至今。	250
Haishan Tidal Power Station	It began construction in 1972, was put into use in 1975, and has been in operation so far.	
岳浦潮汐电站	1970年开始建造，1978年停止运行。	
Yuepu Tidal Power Station	It began construction in 1970, stopped power generation in 1978.	
白沙口潮汐电站	1970年开始建造，1978年投入使用，2010年停止运行。	
Baishakou Tidal Power Station	It began construction in 1970, was put into use in 1978, stopped power generation in 2010.	
果子山潮汐电站	已经投入前期筹备工作阶段，未确定开工时间。	
Guozishan Tidal Power Station	It has been put into the first stage of preparation, and the openning date has not been fixed.	

4-8 主要海上活动船舶
Major Vessels Operating on the Sea

类　别 Type	艘数 （艘） Number of Vessels (unit)	总吨 （万吨位） grt (10 000 t)	净载重量 （万吨） Net Weight Tonnage (10 000 t)	载客量 （客位） Passenger Spaces (seat)	总功率 （千瓦） Total Power (kW)
一 、海洋生产用船 **Vessels for Marine Production**					
海洋渔业船舶 Marine Fishing Vessels	290 566	726.77			15 895 733
远洋渔船 Ocean-going Fishing Vessels	1 587				921 300
海洋油气船舶 Offshore Oil and Gas Vessels	173	87	37	9 557	854 313
钻井平台 Drilling Vessels	39	35.00	10.00	5 292	257 272
物探船 Physical Exploration Vessels	9	3.00	1.00	469	42 601
其 他 Others	125	49.00	26.00	3 796	554 440
海洋运输船舶 Maritime Transport Vessels	12 357	7 788	12 011	189 466	30 644 132
二、海洋科研用船 **Vessels for Marine Scientific Research**					
海洋地质勘探船 Marine Geology Survey Vessels	6	1.18	0.37	415	23 374
海洋调查船 Marine Research Vessels					
中国科学院 Chinese Academy of Sciences	3	0.63	0.20	129	12 161
国家海洋局 State Oceanic Administration	4	2.60	1.35	318	25 012
三、海洋执法用船 **Vessels for Marine Law Enforcement**					
海监船* Marine Monitoring Vessels	366				
渔政执法船 Fishing Monitoring Vessels	520	4.38			226 024

注：*包含船、艇数据。
Note: * includes the data of naval vessels.

4-9 沿海规模以上港口生产用码头泊位
Berths for Productive Use at above Designed Size Seaports

单位：米, 个 (m, number)

港口 Seaport		码头长度 Length of Quay Line	泊位个数 Number of Berths	#万吨级 10 000 Tonnage Class
合　计	**Total**	**617 746**	**4 733**	**1 366**
丹　东	Dandong	5 292	34	17
大　连	Dalian	33 978	198	79
营　口	Yingkou	14 731	69	44
锦　州	Jinzhou	5 375	20	18
秦皇岛	Qinhuangdao	14 750	66	42
黄　骅	Huanghua	5 570	25	19
唐　山	Tangshan	13 449	53	50
天　津	Tianjin	31 366	143	98
烟　台	Yantai	16 140	82	53
威　海	Weihai	3 225	15	10
青　岛	Qingdao	19 500	75	59
日　照	Rizhao	11 989	48	41
上　海	Shanghai	72 742	606	150
连云港	Lianyungang	10 158	53	38
嘉　兴	Jiaxing	6 891	46	23
宁波-舟山	Ningbo-Zhoushan	73 755	625	129
台　州	Taizhou	10 813	172	6
温　州	Wenzhou	16 449	239	15
宁　德	ningde	4 849	48	2

4-9 续表 continued

港 口 Seaport		码头长度 Length of Quay Line	泊位个数 Number of Berths	#万吨级 10 000 Tonnage Class
福 州	Fuzhou	16 480	125	43
莆 田	Putian	4 204	43	5
泉 州	Quanzhou	13 802	104	19
厦 门	Xiamen	23 238	134	60
汕 头	Shantou	9 444	86	18
汕 尾	Shanwei	1 286	14	1
惠 州	Huizhou	6 188	36	18
深 圳	Shenzhen	30 055	160	68
虎 门	humen	12 026	97	21
广 州	Guangzhou	44 398	487	65
中 山	Zhongshan	3 487	61	0
珠 海	Zhuhai	12 864	126	17
江 门	Jiangmen	9 196	144	2
阳 江	Yangjiang	2 232	10	9
茂 名	Maoming	2 333	18	8
湛 江	Zhanjiang	15 757	153	31
北部湾港	Beibuwan	27 136	227	56
海 口	Haikou	5 647	52	10
洋 浦	Yangpu	5 197	29	14
八 所	Basuo	1 754	10	8

4-10 沿海地区星级饭店基本情况
Star-rated Hotels and Occupancies by Coastal Regions

地　区 Region	饭店数（座） Number of Hotels (unit)	客房数（间） Number of Rooms (unit)	床位数（张） Number of Beds (unit)	客房出租率（%） Room Occupancy (%)
合　计 Total	**5 423**	**742 160**	**1 249 923**	**60.00**
天　津 Tianjin	100	16 850	27 462	50.36
河　北 Hebei	395	49 805	91 516	54.58
辽　宁 Liaoning	409	53 050	89 647	59.10
上　海 Shanghai	277	59 878	91 966	55.62
江　苏 Jiangsu	723	90 785	150 568	61.65
浙　江 Zhejiang	844	110 625	187 945	61.62
福　建 Fujian	377	52 008	85 841	62.88
山　东 Shandong	869	100 593	176 268	67.09
广　东 Guangdong	959	140 252	228 244	59.22
广　西 Guangxi	317	40 763	72 208	62.95
海　南 Hainan	153	27 551	48 258	64.95

4-11 沿海地区旅行社数
Number of Travel Agencies by Coastal Regions

单位：家　　(number)

地　区 Region	旅行社总数 Number of Travel Agencies
合　计　Total	**11 962**
天　津　Tianjin	332
河　北　Hebei	1 156
辽　宁　Liaoning	1 116
上　海　Shanghai	1 010
江　苏　Jiangsu	1 891
浙　江　Zhejiang	1 760
福　建　Fujian	702
山　东　Shandong	1 865
广　东　Guangdong	1 376
广　西　Guangxi	459
海　南　Hainan	295

主要统计指标解释

1. 海水养殖面积 是指利用海上、滩涂、陆基进行鱼、甲壳类（虾、蟹）、贝、藻等海水经济动植物的人工养殖的水面面积。在报告期内无论是否全部收获或尚未收获其产品，均应统计在海水养殖面积中。但有些滩涂、水面不投放苗种或投放少量苗种，只进行一般管理的，不统计为养殖面积。

2. 盐田总面积 指盐田占有的全部面积。包括储卤、蒸发、保卤、结晶面积、滩内的沟、壕、池、埝、滩坨等面积及滩外的沟、壕、公路及杂地面积。

3. 生产面积 指直接提供给海盐生产的面积，包括结晶面积、蒸发面积、保卤面积，滩内的沟、壕、池、埝面积及滩坨面积。

4. 年末海盐生产能力 指年末企业生产原盐的全部设备的综合平衡能力。海盐生产露天作业，受天气影响，因而计算生产能力时，成熟滩田按 10 年实际平均单位生产面积产量乘以本年成熟滩田生产面积而得，新滩田按设计能力及滩田成熟程度可能达到的产量计算。

5. 海洋渔业船舶 是指配置机器作为动力的从事海洋渔业生产和辅助渔业生产的船舶。

6. 远洋渔船 按我国远洋渔业项目管理办法在非我国管辖海域（外国专属经济区水域或公海）进行常年或季节性生产的渔船。

7. 泊位个数 是指设有系靠船舶设施，在同一时间内可供靠泊最大吨级船舶的艘数。即可靠泊一艘船舶，则计为一个泊位，余类推。泊位分码头泊位和浮筒泊位。

8. 客房数 指饭店实际可用于接待旅游者的房间数。

9. 床位数 指饭店实际可用于接待旅游者的床位数。

Explanatory Notes on Main Statistical Indicators

1. Mariculture Area refers to the area of the water surface where seawater economic animals and plants, such as crustacean (shrimp, crab), shellfish and algae, are cultivated at sea, on tidal flat and land. Whether or not all the products in the area have been harvested or the products have not been harvested yet in the period covered by the report, the area is included in the Mariculture Area. But some tidal flats and water surfaces where none or a small amount of the young have been released and only general management is carried out are not included in the Mariculture Area.

2. Total Area of Salt Pans refers to the total area covered by salt pans, including the area for brine storage, evaporation, brine preservation, and crystallization, the area of ditches, moats, pondsand banks within the beach as well as beach mounds, and ditches, moats, highway beyond the beach as well as the area of miscellaneous lands.

3. Area of Salt Pan Production refers to the area directly provided for sea-salt productions, including the area for crystallization, evaporation and brine preservation, the area of ditches, moats, ponds and banks within the beach as well as the area of beach mounds.

4. Year-End Capacity of Crude Salt Production refers to the integrated and balanced capacity of all equipment of the enterprise used for crude salt production at the end of the year. As sea salt production is an open-air operation, which is subject to the effect of weather, the production capacity of a matured salt pan is calculated at the productions of the actual average unit production area in ten years times the production area of the matured salt pan in the current year. The production capacity of new salt pans is calculated at the production that may be reached in the light of the designed capacity and the level of maturity of the salt pan.

5. Marine Fishing Vessels refer to the vessels equipped with machines as motive power and going for marine fishery production and auxiliary fishery production.

6. Deep Sea Fishing Vessels refer to the fishing vessels which carry out production all the year round or seasonally in the non-Chinese jurisdictional, sea areas (foreign EEZ or high sea) according to the China Deep-Sea Fishing Projects Management Measures.

7. Number of Berths refers to the spaces equipped with facilities for docking ships and the number of ships of the maximum tonnage that may dock or anchor in them. A space for a ship to dock is counted as one berth and the rest are reasoned out by analogy. Berths are divided into wharf berths and buoy berths.

8. Number of Rooms refers to the number of guest rooms actually used by the hotels receiving tourists.

9. Number of Beds refers to the number of beds actually used by the hotels receiving tourists.

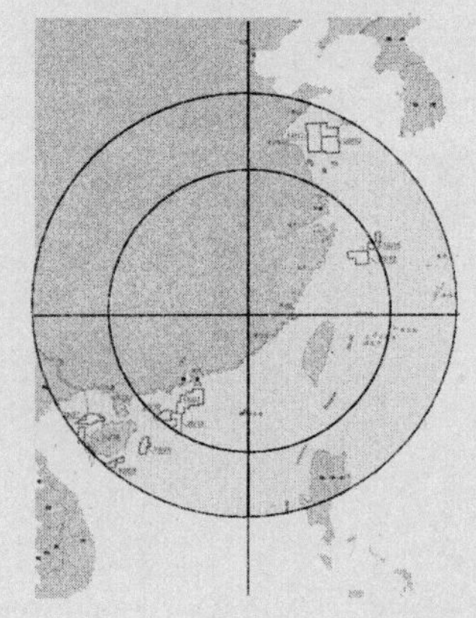

5

涉 海 就 业

Ocean-Related Employment

5-1 全国涉海就业人员情况
National Ocean-Related Employed Personnel

单位：万人 (10 000 persons)

地区 Region	2001	2010①	2011①
合计 Total	**2 107.6**	**3 350.8**	**3 421.7**
天津 Tianjin	106.4	169.2	172.7
河北 Hebei	58.0	92.2	94.2
辽宁 Liaoning	196.0	311.6	318.2
上海 Shanghai	127.5	202.7	207.0
江苏 Jiangsu	116.9	185.9	189.8
浙江 Zhejiang	256.4	407.6	416.3
福建 Fujian	259.7	412.9	421.6
山东 Shandong	319.9	508.6	519.4
广东 Guangdong	505.3	803.4	820.4
广西 Guangxi	68.9	109.5	111.9
海南 Hainan	80.6	128.1	130.9
其他② Others	12.0	19.1	19.5

注：① 2010年和2011年为推算数据；② 其他为非沿海地区涉海就业人员数。

Note: ① The data for 2010 and 2011 are the estimated ones;

② Others refer to the number of ocean-related employed persons in the non-coastal regions.

5-2 全国主要海洋产业就业人员情况
Employed Personnel in the Major Marine Industries throughout the Country

单位：万人 (10 000 persons)

海洋产业 Marine Industry	2001	2010	2011
合计 Total	**719.1**	**1 142.2**	**1 167.5**
海洋渔业及相关产业 Marine Fishery and the Related Industries	348.3	553.2	565.5
海洋石油和天然气业 Offshore Oil and Natural Gas Industry	12.4	19.7	20.1
海滨砂矿业 Beach Placer Industry	1.0	1.6	1.6
海洋盐业 Sea Salt Industry	15.0	23.8	24.4
海洋化工业 Marine Chemical Industry	16.1	25.6	26.1
海洋生物医药业 Marine Biomedicine Industry	0.6	1.0	1.0
海洋电力和海水利用业 Marine Electric Power and Seawater Utilization Industry	0.7	1.1	1.1
海洋船舶工业 Marine Shipbuilding Industry	20.6	32.7	33.4
海洋工程建筑业 Marine Engineering Architecture Industry	38.8	61.6	63.0
海洋交通运输业 Maritime Communications and Transportation Industry	50.8	80.7	82.5
滨海旅游业 Coastal Tourism	78.3	124.4	127.1
其他海洋产业 Other Marine Industries	136.5	216.8	221.6

注：2010年和2011年为推算数据。

Note: The data for 2010 and 2011 are the estimated ones.

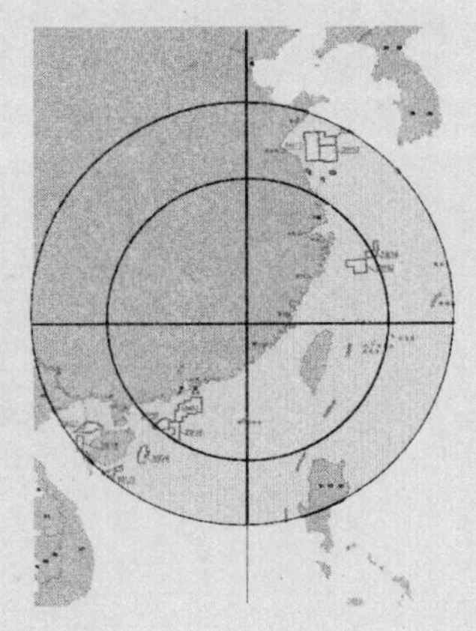

6

海洋科学技术

Marine Science and Technology

6-1 分行业海洋科研机构及人员情况
Marine Scientific Research Institutions and Personnel by Industry

行　业 Industry	机构数（个） Number of Institutions(unit)	从业人员（人） Employed Population(person)
合 计 Total	**179**	**37 445**
海洋基础科学研究 Marine Basic Scientific Research	**102**	**17 005**
海洋自然科学 Marine Natural Science	56	12 869
海洋社会科学 Marine Social Science	4	800
海洋农业科学 Marine Agricultural Science	39	3 247
海洋生物医药 Marine Biomedicine	3	89
海洋工程技术研究 Marine Engineering Technology Research	**65**	**18 624**
海洋化学工程技术 Marine Chemical Engineeing Technology	12	6 333
海洋生物工程技术 Marine Bioengineering Technology	2	214
海洋交通运输工程技术 Marine Communications and Transport Technology	14	3 952

6-1 续表 continued

行 业 Industry	机构数（个） Number of Institutions(unit)	从业人员（人） Employed Population(person)
海洋能源开发技术 Marine Energies Development Technology	4	3 113
海洋环境工程技术 Marine Environmental Engineering Technology	11	975
河口水利工程技术 Eustuarine Water Conservancy Engineering Technology	18	3 183
其他海洋工程技术 Other Marine Engineering Technology	4	854
海洋信息服务业 **Marine Information Service**	**9**	**1 114**
其他海洋信息服务 Other Marine Information Services	9	1 114
海洋技术服务业 **Marine Technological Service Industry**	**3**	**702**
其他海洋专业技术服务 Other Marine Professional and Technological Services	2	121
海洋工程管理服务 Marine Engineering Management Service	1	581

注：机构为县级以上科研机构；行业分类参照《海洋及相关产业分类》标准。

Note: The institutions are the scientific research institutions above the county level;The classification of industries follows the standard *Classification of Marine Industries and the Related Industries*.

6-2 分行业海洋科研机构科技活动人员学历构成
Educational Background Composition of the Personnel Engaged in Scientific and Technological Activities in Marine Scientific Research Institutions by Industry

单位：人 (person)

行　业 Industry	科技活动人员 Personnel Engaged in Scientific Activities	博士 Doctor	硕士 Master	大学生 Graduate	大专生 College Graduate
合　计 Total	**30 642**	**6 252**	**7 945**	**10 137**	**3 564**
海洋基础科学研究 Marine Basic Scientific Research	**14 794**	**4 100**	**3 472**	**4 373**	**1 502**
海洋自然科学 Marine Natural Science	11 302	3 706	2 744	3 020	1 052
海洋社会科学 Marine Social Science	765	89	97	280	85
海洋农业科学 Marine Agricultural Science	2 656	303	613	1 037	356
海洋生物医药 Marine Biomedicine	71	2	18	36	9
海洋工程技术研究 Marine Engineering Technology Research	**14 267**	**2 055**	**3 952**	**5 029**	**1 885**
海洋化学工程技术 Marine Chemical Engineeing Technology	5 261	655	1 139	1 789	873
海洋生物工程技术 Marine Bioengineering Technology	169	8	67	73	19

6-2 续表 continued

行 业 Industry	科技活动人员 Personnel Engaged in Scientific Activities	博士 Doctor	硕士 Master	大学生 Graduate	大专生 College Graduate
海洋交通运输工程技术 Marine Communications and Transport Technology	2 344	84	657	992	384
海洋能源开发技术 Marine Energies Development Technology	2 556	794	811	637	205
海洋环境工程技术 Marine Environmental Engineering Technology	807	40	223	426	85
河口水利工程技术 Eustuarine Water Conservancy Engineering Technology	2 560	404	785	946	278
其他海洋工程技术 Other Marine Engineering Technology	570	70	270	166	41
海洋信息服务业 **Marine Information Service**	**903**	**69**	**269**	**386**	**135**
其他海洋信息服务 Other Marine Information Services	903	69	269	386	135
海洋技术服务业 **Marine Technological Service Industry**	**678**	**28**	**252**	**349**	**42**
其他海洋专业技术服务 Other Marine Professional and Technological Services	108	2	14	69	16
海洋工程管理服务 Marine Engineering Management Service	570	26	238	280	26

6-3 分行业海洋科研机构科技活动人员职称构成
Technical Title Composition of Personnel Engaged in Scientific and Technological Activities in the Marine Scientific Research Institutions by Industry

单位：人 (person)

行 业 Industry	科技活动人员 Personnel Engaged in Scientific Activities	高级职称 Senior	中级职称 Intermediate	初级职称 Primary
合 计 **Total**	**30 642**	**11 774**	**10 015**	**5 357**
海洋基础科学研究 **Marine Basic Scientific Research**	**14 794**	**5 793**	**5 102**	**2 484**
海洋自然科学 Marine Natural Science	11 302	4 623	3 924	1 829
海洋社会科学 Marine Social Science	765	314	239	62
海洋农业科学 Marine Agricultural Science	2 656	841	920	560
海洋生物医药 Marine Biomedicine	71	15	19	33
海洋工程技术研究 **Marine Engineering Technology Research**	**14 267**	**5 433**	**4 403**	**2 529**
海洋化学工程技术 Marine Chemical Engineeing Technology	5 261	1 864	1 650	731
海洋生物工程技术 Marine Bioengineering Technology	169	49	67	53

6-3 续表 continued

行 业 Industry	科技活动人员 Personnel Engaged in Scientific Activities	高级职称 Senior	中级职称 Intermediate	初级职称 Primary
海洋交通运输工程技术 Marine Communications and Transport Technology	2 344	623	675	686
海洋能源开发技术 Marine Energies Development Technology	2 556	1 317	665	322
海洋环境工程技术 Marine Environmental Engineering Technology	807	257	332	170
河口水利工程技术 Eustuarine Water Conservancy Engineering Technology	2 560	1 072	841	445
其他海洋工程技术 Other Marine Engineering Technology	570	251	173	122
海洋信息服务业 **Marine Information Service**	**903**	**327**	**270**	**195**
其他海洋信息服务 Other Marine Information Services	903	327	270	195
海洋技术服务业 **Marine Technological Service Industry**	**678**	**221**	**240**	**149**
其他海洋专业技术服务 Other Marine Professional and Technological Services	108	52	30	19
海洋工程管理服务 Marine Engineering Management Service	570	169	210	130

6-4 分行业海洋科研机构经费收入
Routine Fund Receipts of the Marine Scientific Research Institutions by Industry

单位：万元 (10 000 yuan)

行 业 Industry	经费收入总额 Fund Total	经常费 Routine Fund	科技活动借贷款 Loans in the Scientific and Technological Activities	基本建设中政府投资 Government Investment in the Capital Construction
合 计 Total	**23 221 895**	**21 666 523**	**165 594**	**1 389 778**
海洋基础科学研究 Marine Basic Scientific Research	**10 860 191**	**9 797 308**	**17 594**	**1 045 289**
海洋自然科学 Marine Natural Science	9 319 079	8 329 030	17 594	972 455
海洋社会科学 Marine Social Science	359 761	320 963	0	38 798
海洋农业科学 Marine Agricultural Science	1 163 912	1 129 975	0	33 937
海洋生物医药 Marine Biomedicine	17 439	17 340	0	99
海洋工程技术研究 Marine Technological Service Industry	**11 526 535**	**11 034 094**	**148 000**	**344 441**
海洋化学工程技术 Marine Chemical Engineeing Technology	3 714 639	3 657 793	28 000	28 846
海洋生物工程技术 Marine Bioengineering Technology	154 757	154 757	0	0

6-4 续表 continued

行 业 Industry	经费收入总额 Fund Total	经常费 Routine Fund	科技活动借贷款 Loans in the Scientific and Technological Activities	基本建设中政府投资 Government Investment in the Capital Construction
海洋交通运输工程技术 Marine Communications and Transport Technology	2 457 810	2 079 693	120 000	258 117
海洋能源开发技术 Marine Energies Development Technology	2 735 002	2 732 822	0	2 180
海洋环境工程技术 Marine Environmental Engineering Technology	458 812	458 712	0	100
河口水利工程技术 Eustuarine Water Conservancy Engineering Technology	1 425 201	1 370 003	0	55 198
其他海洋工程技术 Other Marine Engineering Technology	580 314	580 314	0	0
海洋信息服务业 **Marine Information Service**	**481 658**	**481 658**	**0**	0
其他海洋信息服务 Other Marine Information Services	481 658	481 658	0	0
海洋技术服务业 **Marine Technological Service Industry**	**353 511**	**353 463**	**0**	**48**
其他海洋专业技术服务 Other Marine Professional and Technological Services	246 156	246 108	0	48
海洋工程管理服务 Marine Engineering Management Service	107 355	107 355	0	0

6-5 分行业海洋科研机构科技课题情况
Marine Science and Technology Research Projects of the Research Institutions by Industry

单位：项 (item)

行业 Industry	课题数 Number of research subjects	基础研究 Basic Research	应用研究 Applied Research	试验发展 Experimental Development	成果应用 Result Application	科技服务 Scientific and Technological Service
合计 Total	**14 253**	**3 476**	**3 696**	**2 855**	**1 453**	**2 773**
海洋基础科学研究 Marine Basic Scientific Research	**10 497**	**3 440**	**3 246**	**1 520**	**747**	**1 544**
海洋自然科学 Marine Natural Science	8 494	3 260	2 858	936	508	932
海洋社会科学 Marine Social Science	477	4	14	84	22	353
海洋农业科学 Marine Agricultural Science	1 498	176	372	483	208	259
海洋生物医药 Marine Biomedicine	28	0	2	17	9	0
海洋工程技术研究 Marine Engineering Technology Research	**3 571**	**36**	**444**	**1 291**	**697**	**1 103**
海洋化学工程技术 Marine Chemical Engineeing Technology	840	7	119	480	168	66
海洋生物工程技术 Marine Bioengineering Technology	17	0	0	0	8	9

6-5 续表 continued

行 业 Industry	课题数 Number of research subjects	基础研究 Basic Research	应用研究 Applied Research	试验发展 Experimental Development	成果应用 Result Application	科技服务 Scientific and Technological Service
海洋交通运输工程技术 Marine Communications and Transport Technology	570	3	31	113	223	200
海洋能源开发技术 Marine Energies Development Technology	722	6	151	354	52	159
海洋环境工程技术 Marine Environmental Engineering Technology	215	0	15	57	38	105
河口水利工程技术 Eustuarine Water Conservancy Engineering	1 135	18	122	259	187	549
其他海洋工程技术 Other Marine Engineering Technology	72	2	6	28	21	15
海洋信息服务业 Marine Information Service	**122**	**0**	**5**	**24**	**2**	**91**
其他海洋信息服务 Other Marine Information Services	122	0	5	24	2	91
海洋技术服务业 Marine Technological Service Industry	**63**	**0**	**1**	**20**	**7**	**35**
其他海洋专业技术服务 Other Marine Professional and Technological Services	16	0	1	10	5	0
海洋工程管理服务 Marine Engineering Management Service	47	0	0	10	2	35

6-6 分行业海洋科研机构科技论著情况
Marine Scientific and Technological Works of the Research Institutions by Industry

行 业 Industry	发表科技论文（篇） Scientific Theses Published (piece)	#国外发表 Published Abroad	出版科技著作（种） Scientific and Technological Works Published (kind)
合 计 Total	**15 547**	**4 169**	**278**
海洋基础科学研究 Marine Basic Scientific Research	**10 431**	**3 477**	**183**
海洋自然科学 Marine Natural Science	8 081	3 238	100
海洋社会科学 Marine Social Science	753	46	53
海洋农业科学 Marine Agricultural Science	1 582	193	30
海洋生物医药 Marine Biomedicine	15	0	0
海洋工程技术研究 Marine Engineering Technology esearch	**4 745**	**672**	**90**
海洋化学工程技术 Marine Chemical Engineeing Technology	947	112	7
海洋生物工程技术 Marine Bioengineering Technology	3	0	0

6-6 续表 continued

行　业 Industry	发表科技论文（篇） Scientific Theses Published (piece)	#国外发表 Published Abroad	出版科技著作（种） Scientific and Technological Works Published (kind)
海洋交通运输工程技术 Marine Communications and Transport Technology	656	64	2
海洋能源开发技术 Marine Energies Development Technology	1 166	177	18
海洋环境工程技术 Marine Environmental Engineering Technology	144	12	3
河口水利工程技术 Eustuarine Water Conservancy Engineering	1 371	224	46
其他海洋工程技术 Other Marine Engineering Technology	458	83	14
海洋信息服务业 **Marine Information Service**	**290**	**11**	**5**
其他海洋信息服务 Other Marine Information Services	290	11	5
海洋技术服务业 **Marine Technological Service Industry**	**81**	**9**	**0**
其他海洋专业技术服务 Other Marine Professional and Technological Services	21	2	0
海洋工程管理服务 Marine Engineering Management Service	60	7	0

6-7 分行业科研机构科技专利情况
Marine Scientific and Technological Patents of the Research Institutions by Industry

单位：件 (number)

行　业 Industry	专利申请受理数 Number of Patent Applications Accepted	#发明专利 Patents for Discoveries	专利授权数 Number of Patents Granted	#发明专利 Patents for Discoveries	拥有发明专利总数 Total Number of Patents for Discoveries
合　计 Total	**4 412**	**3 667**	**2 034**	**1 355**	**8 009**
海洋基础科学研究 Marine Basic Scientific Research	**1 305**	**979**	**924**	**603**	**2 251**
海洋自然科学 Marine Natural Science	1 066	821	676	458	1 941
海洋社会科学 Marine Social Science	0	0	1	1	5
海洋农业科学 Marine Agricultural Science	237	156	246	143	294
海洋生物医药 Marine Biomedicine	2	2	1	1	11
海洋工程技术研究 Marine Engineering Technology Research	**3 060**	**2 672**	**1 078**	**742**	**5 731**
海洋化学工程技术 Marine Chemical Engineeing Technology	2 471	2 358	726	633	5 220
海洋生物工程技术 Marine Bioengineering Technology	0	0	0	0	0

6-7 续表 continued

行　业 Industry	专利申请受理数 Number of Patent Applications Accepted	#发明专利 Patents for Discoveries	专利授权数 Number of Patents Granted	#发明专利 Patents for Discoveries	拥有发明专利总数 Total Number of Patents for Discoveries
海洋交通运输工程技术 Marine Communications and Transport Technology	69	33	44	17	56
海洋能源开发技术 Marine Energies Development Technology	247	142	160	46	283
海洋环境工程技术 Marine Environmental Engineering Technology	16	6	12	3	4
河口水利工程技术 Eustuarine Water Conservancy Engineering Technology	91	43	84	25	120
其他海洋工程技术 Other Marine Engineering Technology	166	90	52	18	48
海洋信息服务业 Marine Information Service	**25**	**2**	**17**	**0**	**0**
其他海洋信息服务 Other Marine Information Services	25	2	17	0	0
海洋技术服务业 Marine Technological Service Industry	**22**	**14**	**15**	**10**	**27**
其他海洋专业技术服务 Other Marine Professional and Technological Services	20	14	14	10	23
海洋工程管理服务 Marine Engineering Management Service	2	0	1	0	4

6-8 分行业海洋科研机构R&D情况
R&D in the Marine Scientific Research Institutions by Professions

行　业 Industry	R&D人员 （人） R&D Personnel (persons)	R&D经费内部支出 （千元） R&D Internal Expenditure (1 000 yuan)	R&D课题数 （项） Number of R&D Projects (item)
合　计 Total	**25 077**	**10 913 150**	**10 027**
海洋基础科学研究 Marine Basic Scientific Research	**17 531**	**6 747 840**	**8 206**
海洋自然科学 Marine Natural Science	15 139	6169 118	7 054
海洋社会科学 Marine Social Science	550	67 053	102
海洋农业科学 Marine Agricultural Science	1 799	503 369	1 031
海洋生物医药 Marine Biomedicine	43	8 300	19
海洋工程技术研究 Marine Engineering Technology Research	**7 052**	**3 993 925**	**1 771**
海洋化学工程技术 Marine Chemical Engineeing Technology	3 421	1863 510	606
海洋生物工程技术 Marine Bioengineering Technology	0	0	0

6-8 续表 continued

行 业 Industry	R&D人员 （人）	R&D经费内部支出 （千元）	R&D课题数 （个）
海洋交通运输工程技术 Marine Communications and Transport Technology	946	252 431	147
海洋能源开发技术 Marine Energies Development Technology	1 312	1408 802	511
海洋环境工程技术 Marine Environmental Engineering Technology	285	84 401	72
河口水利工程技术 Eustuarine Water Conservancy Engineering Technology	916	300 891	399
其他海洋工程技术 Other Marine Engineering Technology	172	83 890	36
海洋信息服务业 Marine Information Service	**326**	**122 127**	**29**
其他海洋信息服务 Other Marine Information Services	326	122 127	29
海洋技术服务业 Marine Technological Service Industry	**168**	**49 258**	**21**
其他海洋专业技术服务 Other Marine Professional and Technological Services	62	27 568	11
海洋工程管理服务 Marine Engineering Management Service	106	21 690	10

6-9 分地区海洋科研机构及人员情况
Marine Scientific Research Institutions and Personnel by Regions

地 区 Region	机构数（个） Number of Institutions (unit)	从业人员（人） Employed Population (person)
合 计 **Total**	**179**	**37 445**
北 京 Beijing	25	13 704
天 津 Tianjin	14	2 586
河 北 Hebei	5	554
辽 宁 Liaoning	17	2 118
上 海 Shanghai	15	3 542
江 苏 Jiangsu	11	3 295
浙 江 Zhejiang	17	1 614
福 建 Fujian	12	1 023
山 东 Shandong	22	3 719
广 东 Guangdong	25	3 088
广 西 Guangxi	9	466
海 南 Hainan	3	185
其 他 Other	4	1 551

6-10 分地区海洋科研机构科技活动人员学历构成
Educational Background Composition of the Personnel Engaged in Scientific and Technological Activities in Marine Scientific Research Institutions by Regions

单位：人 (person)

地 区 Region	科技活动人员 Personnel Engaged in Scientifical Activities	博士 Doctor	硕士 Master	大学生 Graduate	大专生 College Graduate
合 计 Total	**30 642**	**6 252**	**7 945**	**10 137**	**3 564**
北 京 Beijing	11 949	3 276	3 074	3 310	1 221
天 津 Tianjin	2 056	125	605	937	225
河 北 Hebei	535	25	113	259	78
辽 宁 Liaoning	1 601	116	408	622	235
上 海 Shanghai	3 011	451	855	1 113	378
江 苏 Jiangsu	1 943	281	454	671	320
浙 江 Zhejiang	1 336	113	413	597	173
福 建 Fujian	968	96	265	409	97
山 东 Shandong	3 049	653	737	999	398
广 东 Guangdong	2 564	639	612	725	279
广 西 Guangxi	365	13	68	200	65
海 南 Hainan	143	5	34	50	24
其 他 Other	1 122	459	307	245	71

6-11　分地区海洋科研机构科技活动人员职称构成
Technical Title Composition of Personel Engaged in Marine Scientific Research Institutions by Regions

单位：人　　(person)

地　区 Region	科技活动人员 Personnel Engaged in Scientifical Activities	高级职称 Senior	中级职称 Intermediate	初级职称 Primary
合　计 Total	**30 642**	**11 774**	**10 015**	**5 357**
北　京 Beijing	11 949	5 186	3 904	1 526
天　津 Tianjin	2 056	711	685	436
河　北 Hebei	535	209	129	33
辽　宁 Liaoning	1 601	531	511	244
上　海 Shanghai	3 011	1 005	1 052	671
江　苏 Jiangsu	1 943	733	471	382
浙　江 Zhejiang	1 336	485	449	243
福　建 Fujian	968	286	340	241
山　东 Shandong	3 049	1 113	1 083	695
广　东 Guangdong	2 564	878	828	525
广　西 Guangxi	365	67	147	125
海　南 Hainan	143	13	33	58
其　他 Other	1 122	557	383	178

6-12 分地区海洋科研机构经费收入
Routine Fund Receipts of Marine Scientific Research Institutions by Regions

单位：万元 (10 000 yuan)

地 区 Region	经费收入总额 Fund Total	经常费 Routine Fund	科技活动借贷款 Loans in the Scientific and Technological Activities	基本建设中政府投资 Government Investment in the Capital Construction
合 计 Total	**23 221 895**	**21 666 523**	**165 594**	**1 389 778**
北 京 Beijing	9 118 227	8 785 241	15 000	317 986
天 津 Tianjin	1 593 334	1 376 177	0	217 157
河 北 Hebei	131 596	121 401	0	10 195
辽 宁 Liaoning	1 051 461	1 050 861	0	600
上 海 Shanghai	2 674 753	2 544 626	0	130 127
江 苏 Jiangsu	1 726 401	1 547 793	120 000	58 608
浙 江 Zhejiang	1 114 879	1 113 966	0	913
福 建 Fujian	527 127	526 262	0	865
山 东 Shandong	2 545 958	1 991 871	25 000	529 087
广 东 Guangdong	1 744 704	1 670 061	3 000	71 643
广 西 Guangxi	83 236	83 236	0	0
海 南 Hainan	52 398	49 851	0	2 547
其 他 Other	857 821	805 177	2 594	50 050

6-13 分地区海洋科研机构科技课题情况
Marine Science and Technology Research Projects of the Research Institutions by Regions

单位：项 (item)

地 区 Region	课题数 Number of research subjects	基础研究 Basic Research	应用研究 Applied Research	试验发展 Experimental Development	成果应用 Result Application	科技服务 Scientific and Technological Service
合 计 Total	**14 253**	**3 476**	**3 696**	**2 855**	**1 453**	**2 773**
北 京 Beijing	4 897	1 302	1 126	931	382	1 156
天 津 Tianjin	536	0	59	239	74	164
河 北 Hebei	67	4	17	11	15	20
辽 宁 Liaoning	290	0	21	149	82	38
上 海 Shanghai	1 094	132	255	299	121	287
江 苏 Jiangsu	1 718	85	557	394	467	215
浙 江 Zhejiang	440	46	99	69	53	173
福 建 Fujian	627	208	163	58	94	104
山 东 Shandong	1 477	421	573	286	57	140
广 东 Guangdong	1 929	579	629	250	47	424
广 西 Guangxi	104	4	24	55	10	11
海 南 Hainan	84	0	9	24	44	7
其 他 Other	990	695	164	90	7	34

6-14 分地区海洋科研机构科技论著情况
Marine Scientific and Technological Works of the Research Institutions by Regions

地 区 Region	发表科技论文（篇） Scientific Theses Published (piece)	#国外发表 Published Abroad	出版科技著作（种） Scientific and Technological Works Published (kind)
合 计 Total	**15 547**	**4 169**	**278**
北 京 Beijing	5 953	1 763	106
天 津 Tianjin	765	101	14
河 北 Hebei	555	0	38
辽 宁 Liaoning	446	35	0
上 海 Shanghai	1 103	282	17
江 苏 Jiangsu	1 005	231	22
浙 江 Zhejiang	497	95	13
福 建 Fujian	406	100	11
山 东 Shandong	1 879	568	28
广 东 Guangdong	1 552	469	14
广 西 Guangxi	142	2	0
海 南 Hainan	56	38	1
其 他 Other	1 188	485	14

6-15 分地区海洋科研机构科技专利情况
Marine Scientific and Technological Patents of the Research Institutions by Regions

单位：件 (number)

地 区 Region	专利申请受理数 Number of Patent Applications Accepted	#发明专利 Patents for Discoveries	专利授权数 Number of Patents Granted	#发明专利 Patents for Discoveries	拥有发明专利总数 Total Number of Patents for Discoveries
合 计 Total	**4 412**	**3 667**	**2 034**	**1 355**	**8 009**
北 京 Beijing	1 974	1 772	812	614	4 277
天 津 Tianjin	111	59	67	13	100
河 北 Hebei	6	6	1	1	8
辽 宁 Liaoning	546	493	168	110	1 055
上 海 Shanghai	831	734	400	293	1 204
江 苏 Jiangsu	133	51	63	19	55
浙 江 Zhejiang	67	43	41	26	61
福 建 Fujian	23	6	20	15	105
山 东 Shandong	232	188	213	142	432
广 东 Guangdong	278	205	156	91	544
广 西 Guangxi	33	7	19	2	2
海 南 Hainan	2	2	5	2	2
其 他 Other	176	101	69	27	164

6-16 分地区海洋科研机构R&D情况
R&D in the Marine Scientific Research Institutions by Regions

行　业 Industry	R&D人员 （人） R&D Personnel (persons)	R&D经费内部支出 （千元） R&D Internal Expenditure (1 000 yuan)	R&D课题数 （项） Number of R&D Projects (item)
合　计 **Total**	**25 077**	**10913 150**	**10 027**
北　京 Beijing	10 039	4768 236	3 359
天　津 Tianjin	1 302	522 200	298
河　北 Hebei	258	33 325	32
辽　宁 Liaoning	696	322 016	170
上　海 Shanghai	2 392	1284 748	686
江　苏 Jiangsu	1 536	583 736	1 036
浙　江 Zhejiang	614	319 300	214
福　建 Fujian	475	220 009	429
山　东 Shandong	2 769	1518 322	1 280
广　东 Guangdong	3 128	858 228	1 458
广　西 Guangxi	171	26 669	83
海　南 Hainan	35	8 930	33
其　他 Other	1 662	447 431	949

主要统计指标解释

1. 海洋科研机构　指有明确的研究方向和任务，有一定水平的学术带头人和一定数量、质量的研究人员，有开展研究工作的基本条件，长期有组织地从事海洋研究与开发活动的机构。

2. 从业人员　指由本机构年末直接组织安排工作并支付工资的各类人员总数。包括固定职工、国家有编制的合同制职工、招聘人员和返聘的离退休人员。不包括离退休人员、停薪留职人员。

3. 从事科技活动人员　指从业人员中的科技管理人员、课题活动人员和科技服务人员。

4. 高级职称　指研究员、副研究员；教授、副教授；高级工程师；高级农艺师；正、副主任医（药、护、技)师；高级实验师；高级统计师；高级经济师；高级会计师；编审(正、副编审)；译审(正、副译审)、高级(主任)记者；正、副研究馆员等。

5. 中级职称　指助理研究员；讲师；工程师；农艺师；主治医(药、护、技)师；实验师；统计师；经济师；会计师；编辑；翻译；记者；馆员等。

6. 初级职称　指研究实习员；助教；助理工程师、技术员；助理农艺师、农业技术员；医(药、护、技)师、医(药、护、技)士；助理实验师、实验员；助理统计师、统计员；助理经济师；助理会计师、会计员；助理编辑、见习编辑；助理翻译；助理记者；助理馆员、管理员等。

7. 科技经费筹集额　指从各种渠道筹集到的计划用于本单位科技活动的经费，不论来源渠道如何。

8. 政府资金　指由各级政府部门直接拨款或企事业单位利用政府资金委托本机构从事科学技术活动所获得的收入。

9. 生产经营活动收入　指本机构在科研、技术等专业业务活动以外开展非独立核算的经营活动取得的收入，包括产品（商品）销售收入、经营服务收入、工程承包收入、租赁收入和其他经营收入。

10. 其他收入　指开展科技活动与生产经营活动以外的各项活动的收入，包括：用于离退休人员的政府拨款。

11. 非科技活动借贷款　指本机构为开展非科技活动从各种渠道获得的各类借、贷款。不论偿还形式、期限和数额如何，均按当年获得的借、贷款额填报。不包括基本建设贷款。

12. 基础研究　为获得新知识而进行的独创性研究。其目的是揭示观察到的现象和事实的基本原理和规律，而不以任何特定的实际应用为目的。

13. 应用研究　为获得新的科学技术知识而进行的独创性研究。它主要针对某一特定的实际应用目的。应用研究通常是为了确定基础研究成果或知识的可能的用途，或是为达到某一具体的、预定的实际目的确定新的方法(原理性)或途径。

14. 试验发展　利用从研究或实际经验获得的知识，为生产新的材料、产品和装置，建立新的工艺和系统，以及对已生产或建立的上述各项进行实质性的改进而进行的系统性工作。

15. 成果应用　为解决 R&D 活动阶段产生的新产品、新装置、新工艺、新技术、新方法、新系统和服务等能投入生产或在实际应用中所存在的技术问题而进行的系统性活动。它不具有创新成分。此类活动包括为达到生产目的而进行的定型设计和试制以及为扩大新产品的生产规模和探索新方法、新技术、新工艺等的应用领域而进行的适应性试验。

16. 科技服务 与科学研究与实验发展有关，并有助于科学技术知识的产生、传播和应用的活动。包括为扩大科技成果的使用范围而进行的示范性推广工作；为用户提供科技信息和文献服务的系统性工作；为用户提供可行性报告、技术方案、建议及进行技术论证等技术咨询工作；自然、生物现象的日常观测、监测，资源的考查和勘探；有关社会、人文、经济现象的通用资料的收集，如统计、市场调查等以及这些资料的常规分析与整理；为社会和公众提供的测试、标准化、计量、计算、质量控制和专利服务，不包括工商企业为进行正常生产而开展的上述活动。

17. 生产性活动 由于业务特殊的工艺设备条件，或掌握某种技术专长或诀窍，所进行的小量非常规生产。

18. 科技论文 在全国性学报或学术刊物上、省部属大专院校对外正式发行的学报或学术刊物上发表的论文以及向国外发表的论文。

19. 科技著作 经过正式出版部门编印出版的科技专著、大专院校教科书、科普著作。

20. 专利申请受理数 当年本单位向专利管理部门提出申请并被受理的职务专利申请件数。

21. 专利授权数 当年由专利管理部门授予本单位专利权的职务专利件数。

Explanatory Notes on Main Statistical Indicators

1. Marine Scientific Research Institution refers to the institution which has definite research orientations and tasks, high-level academic leading personnel and fair-sized, qualified research personnel, and basic conditions for research work and which is engaged for a long time in the marine research and development activities in an organized way.

2. Employees refer to the total number of personnel of various kinds employed and paid by the institution at the end of the year, including fixed employees, contract workers of staff belonging to the state authorized staff , recruited personnel, reemployed retired personnel, but not including the retired and the personnel on leave with pay suspension.

3. Personnel Engaged in Scientific and Technological Activities refers to the personnel for scientific and Technological management, personnel engaged in the activities of research topics and scientific and technological service personnel.

4. Senior Technical Title refer to research scientist, associate research scientist; professor, associate professor; senior engineer; senior agronomist; professor-rank and associate professor-rank doctor (pharmacists, nurses and technicians); senior laboratory technician; senior statisticians; senior economic engineer; chief accountant; senior editor (professor and associate professor ranks); senior translator (professor and associate professor ranks); senior journalist; research librarian (professor and associate professor ranks), etc.

5. Intermediate Technical Title refer to assistant research scientist; lecturer; engineer; agronomist; lecturer-rank doctor (pharmacist, nurse and technician); laboratory technician; lecturer-rank statistician;

economic engineer; accountant; editor; translator; journalist; librarian, etc.

6. Primary Technical Title refers to trainee researcher; assistant; assistant engineer, technician, assistant agronomist, agricultural technician; assistant-rank doctor (pharmacist, nurse and technician); assistant laboratory technician; assistant statistician; assistant economic engineer, assistant accountant; assistant editor, editor on probation; assistant translator; assistant journalist; assistant research librarian, librarian ,etc.

7. Amount of Scientific and Technological Funds Raised refers to the funds raised through all channels planned to be used as funds for the scientific and technological activities in the institution regardless of their source and channels.

8. Funds from government refers to the direct appropriations by the government departments at all levels or the earnings from conducting scientific and technological activities entrusted to the institution by enterprises as institutions by earning the funds from government.

9. Earnings from Production as Business Activities refer to the incomes obtained from the non-independent accounting business activities carried out by the institution beyond the scientific research, technological and professional activities, including these from sale of products(goods), business and service, contracted projects, leasing and other business.

10. Other Incomes refer to those from the various activities conducted other than scientific and technological activities and production and business activities, including the government appropriations for retired personnel.

11. Loan for Non-scientific as Technological Activities refers to the various types loan obtained by the institutions through all channels for carrying out non-scientific and technological activities, not including the load for capital construction. The loan, irrespective of its form of reimbursement, term and amount is filled in a form and submitted to the authorities as the amount acquired in the current year.

12. Basic Research refers to the original research to acquire new knowledge. It is aimed at revealing the basic principles and laws of the phenomena and facts observed, but not at any specific practical applications.

13. Applied Research refers to the original research to acquire new scientific and technological knowledge. It mainly serves the purpose of a particular practical application. The purpose of applied research is usually to define the potential uses of the research finds or knowledge obtained from basic research or to identify new methods (principles) or ways to reach a specific and predetermined goal.

14. Experimental Development refers to the systematic work carried out to establish new technologies and systems for producing new materials, products and equipment by using the knowledge obtained from research or practical experience, or to make substantial improvement of the above-mentioned which have been produced or established.

15. Result Application refers to the systematic activities carried out to solve the technical problems that might crop up in the production or practical application of the new products, devices, technologies, techniques, methods, systems and service occurring in the course of R&D activities.

They do not bring forth new ideas. Such activities include the finalizing design and trial-production for the purpose of production as well as the adaptive tests to expand the production scale of new products and the application areas of new methods, techniques and technologies.

16. Scientific and Technological Service refers to the activities that are associated with the scientific research and experimental development, and contribute to the generation, dissemination and application of scientific and technological knowledge, which include the demonstrative work of popularization to enlarge the use scope of scientific and technological achievements; the systematic work of providing the users with scientific and technological information and literature service; the technical consultation work of providing users with feasibility reports, technical schemes and recommendations and carrying out technical demonstration; routine observation and monitoring of natural and biological phenomena, and the survey and exploration of resources; collection of universal data on the appropriate social, cultural and economic phenomena, such as statistics and market survey, as well as the routine analysis and sorting-out of these data; the provision for the society and the public of such service as testing, standardization, computation, quality control and patent, but not including the type of the above-mentioned activities carried out by industrial and commercial enterprises for the purpose of normal production.

17. Productive Activity refers to the small-scale and non-conventional productions due to the presence of special technologies and equipment or mastery of a particular technical expertise or secret of success.

18. Scientific Treatises refer to the theses published in the national journals or academic publications, those officially issued journals or academic publications by universities and colleges under provinces or ministries as well as theses published abroad.

19. Scientific and Technological Works refer to the scientific and technological monographs, textbooks for universities and colleges and popular science books published by the official publishing houses.

20. Number of Patents Applied and Accepted refers to the number of professional patent applications of the unit to the patent administrative department and accepted by it in the year.

21. Number of Patents Granted refers to the number of the professional patents granted to the unit by the patent administrative department in the year.

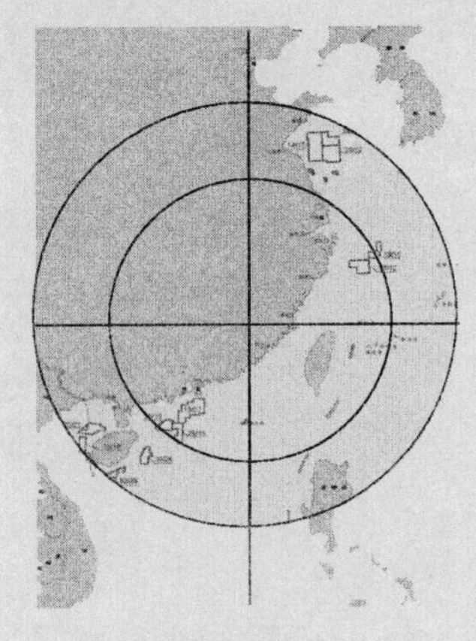

7

海 洋 教 育

Marine Education

7-1 全国各海洋专业博士研究生情况
Doctoral Students from Marine Specialities

专 业 Speciality	专业点数（个） Number of Speciality Agencies (unit)	学生数（人） Students (person)			
		毕业生 Graduated	招 生 Entrants	在校生 Enrollment	毕业班学生 Graduation
合 计 **Total**	**125**	**601**	**819**	**3 584**	**1 780**
物理海洋学 Physical Oceanography	5	35	62	233	119
海洋化学 Marine Chemistry	5	23	32	131	77
海洋生物学 Marine Biology	9	76	71	277	128
海洋地质 Marine Geology	9	35	53	232	127
海洋科学类新专业 New Speciality of Marine Science	3	63	82	299	150
港口海岸及近海工程 Coastal Harbour and Offshore Engineering	10	31	58	302	128
船舶与海洋结构物设计制造 Ship and Marine Structures Design and Manufacture	10	50	66	351	180
轮机工程 Turbine Engineering	9	25	48	258	80

7-1 续表 continued

专 业 Speciality	专业点数（个） Number of Speciality Agencies (unit)	学生数（人） Students (person)			
		毕业生 Graduated	招 生 Entrants	在校生 Enrollment	毕业班学生 Graduation
水声工程 Hydroacoustic Engineering	7	27	32	143	83
船舶与海洋工程新专业 New Speciality of Shipand Marine Engineering	1	0	3	18	12
捕捞学 Science of Fishing	2	0	1	14	10
水产品加工及贮藏工程 Aquatic Product Sprocessing and Storing Engineering	8	7	14	50	24
水产新专业 New Specialities of Aquaculture	3	8	11	53	34
水产养殖 Aquaculture	7	49	57	206	88
水生生物学 Hydrobiology	18	78	94	360	165
水文学及水资源 Hydrology and Water Resource	16	87	125	594	338
渔业资源 Fishery Resource	3	7	10	63	37

7-2 全国各海洋专业硕士研究生情况
Postgraduate Students from Marine Specialities

专 业 Speciality	专业点数（个） Number of Speciality Agencies (unit)	学生数（人）Students (person)			
		毕业生 Graduated	招 生 Entrants	在校生 Enrollment	毕业班学生 Graduation
合 计 Total	**294**	**3 034**	**3 469**	**10 354**	**3 564**
物理海洋学 Physical Oceanography	10	77	119	320	95
海洋化学 Marine Chemistry	14	99	104	301	102
海洋生物学 Marine Biology	25	291	412	1 222	413
海洋地质 Marine Geology	17	112	153	428	124
海洋科学类新专业 New Speciality of Marine Science	3	41	68	207	68
港口海岸及近海工程 Coastal Harbour and Offshore Engineering	22	277	310	934	327
船舶与海洋结构物设计制造 Ship and Marine Structures Design and Manufacture	17	389	362	1 133	421
轮机工程 Turbine Engineering	15	252	255	788	314

7-2 续表 continued

专 业 Speciality	专业点数（个） Number of Speciality Agencies (unit)	学生数（人）Students (person)			
		毕业生 Graduated	招 生 Entrants	在校生 Enrollment	毕业班学生 Graduation
水声工程 Hydroacoustic Engineering	11	131	104	384	149
船舶与海洋工程新专业 New Specialities in Shipping and Marine Engineering	6	27	25	72	25
捕捞学 Science of Fishing	4	23	33	97	36
航空、航天与航海医学 Aeronautical, Aerospace and Nautical Medicine	3	13	16	46	15
水产品加工及贮藏工程 Aquatic Products Processing and Storing Engineering	20	85	99	277	86
水产新专业 New Specialities of Aquaculture	3	19	17	51	20
水产养殖 Aquaculture	29	374	449	1 323	471
水生生物学 Hydrobiology	37	329	390	1 114	343
水文学及水资源 Hydrology and Water Resource	50	419	461	1 405	476
渔业资源 Fishery Resource	8	76	92	252	79

7-3 全国普通高等教育各海洋专业本科学生情况
Undergraduates and Students from Colleges for Professional Training from Marine Specialities in the Ordinary National Higher Education

专 业 Speciality	专业点数（个） Number of Speciality Agencies (unit)	学生数（人）Students (person)			
		毕业生 Graduated	招 生 Entrants	在校生 Enrollment	毕业班学生 Graduation
合 计 Total	**201**	**12 543**	**15 257**	**58 646**	**14 242**
海洋科学 Marine Science	21	744	815	3 262	770
海洋技术 Marine Technology	15	504	538	2 162	545
海洋管理 Marine Management	4	36	78	345	100
海洋生物资源与环境 Marine Living Resources and Environment	5	132	144	531	142
港口海岸及治河工程 Coastal Harbour and River Control Engineering	1	51	0	64	64
港口航道与海岸工程 Harbour Channel and Coastal Engineering	22	1 194	1 555	5 960	1 458
水资源与海洋工程 Water Resource and Ocean Engineering	2	26	27	65	19
航海技术 Nautical Technology	15	1 826	2 501	9 458	2 381
轮机工程 Turbine Engineering	20	2 368	2 820	11 003	2 632
海事管理 Maritime Affairs Manegernent	2	127	128	488	125
船舶与海洋工程 Ship and Marine Engineering	28	2 649	2 748	11 971	3 126
海洋工程类新专业 New Speciality of Marine Engineering	3	0	205	571	67
水产养殖学 Aquaculture	46	2 431	2 771	9 985	2 269
海洋渔业科学与技术 Science and Technology of Marine Fishery	9	277	432	1 539	335
水族科学与技术 Science and Technology of Aquatic Animals	6	137	307	936	209
水产类新专业 New Speciality in Aquatic Products	2	41	188	306	0

7-4 全国普通高等教育各海洋专业专科学生情况
Undergraduates and Students from Colleges for Professional Training from Marine Specialities in the Ordinary National Higher Education

专 业 Speciality	专业点数（个） Number of Speciality Agencies (unit)	学生数（人）Students (person)			
		毕业生 Graduated	招 生 Entrants	在校生 Enrollment	毕业班学生 Graduation
合 计 Total	**448**	**37 515**	**33 845**	**106 581**	**36 608**
港口工程技术 Harbour Engineering	6	310	446	1 323	407
港口业务管理 Harbour Business Management	19	865	1 149	3 145	990
港口与航运管理 Harbour and Shipping Management	6	277	367	911	243
港口机械应用技术 Applied Harbour Machinery Technology	2	296	215	613	162
港口物流设备与自动控制 Harbour Logistics Equipment and Automatic Control	20	1 544	1 042	3 409	1 280
集装箱运输管理 Container Transportation Management	18	1 183	1 010	3 192	1 101
国际航运业务管理 International Shipping Business Management	29	1 761	2 039	5 791	1 900
海关管理 Customs Management	4	292	16	234	204
报关与国际货运 Customs Clearing and International Freight Transport	189	17 046	13 000	44 788	16 159
轮机工程技术 Engine Engineering	44	5 957	6 175	18 520	5 920
船舶检验 Ship Inspection	6	112	189	451	163
船舶舾装 Ship Equipment and Installations	5	300	344	975	335

7-4 续表 continued

专 业 Speciality	专业点数（个） Number of Speciality Agencies (unit)	学生数（人）Students (person)			
		毕业生 Graduated	招 生 Entrants	在校生 Enrollment	毕业班学生 Graduation
船艇动力管理 Ship and Boat Power Equipment Management	1	39	48	128	80
船舶工程技术 Ship Engineering	35	5 386	5 427	16 002	5 296
船机制造与维修 Ship Engines Manufacturing and Maintenance	7	136	492	1 008	202
水运管理 Water Transport Management	3	95	147	323	54
水信息技术 Hydrological Information Technology	2	37	17	94	19
水环境监测与保护 Water Environmental Monitoring and Protection	7	285	205	766	273
水政水资源管理 Water Resources Management by Water Administration	2	44	95	187	44
水产养殖技术 Aquaculture Technology	31	1 154	1 133	3 544	1 304
渔业综合技术 Aquatic animals and plants conservation	3	41	12	185	108
水生动植物保护 Integrated fishery technology	2	81	17	158	75
水上运输类新专业 New Specialities in Waterborne Communications	1	0	0	1	1
港口航道与治河工程 Harbour, channel and river harnessing engineering	6	274	260	833	288

7-5 全国成人高等教育各海洋专业本科学生情况
Students from Marine Specialities in the National Adult Higher Education

专 业 Speciality	专业点数（个） Number of Speciality Agencies (unit)	学生数（人）Students (person)			
		毕业生 Graduated	招 生 Entrants	在校生 Enrollment	毕业班学生 Graduation
合 计 Total	**48**	**1 480**	**2 303**	**5 579**	**2 404**
海洋科学 Marine Science	1	20	0	2	0
海洋技术 Marine Technology	1	0	0	103	0
港口航道与海岸工程 Harbour Channel and Coastal Engineering	3	42	58	127	44
航海技术 Navigation Technology	7	171	324	760	248
轮机工程 Turbine Engineering	7	190	286	680	329
海事管理 Maritime Affairs Management	2	33	3	59	64
船舶与海洋工程 Ship and Marine Engineering	13	788	1 470	3 369	1 544
水产养殖学 Aquaculture	13	226	162	479	175
海洋渔业科学与技术 Science and Technology of Marine Fishery	1	10	0	0	0

7-6 全国成人高等教育各海洋专业专科学生情况
Students from Marine Specialities in the National Adult Higher Education

专业 Speciality	专业点数（个） Number of Speciality Agencies (unit)	学生数（人）Students (person)			
		毕业生 Graduated	招生 Entrants	在校生 Enrollment	毕业班学生 Graduation
合计 Total	**130**	**5 489**	**7 596**	**19 575**	**7 241**
港口业务管理 Harbour Business Management	4	19	2	82	62
港口与航运管理 Harbour and Shipping Management	2	12	0	15	14
国际航运业务管理 International Shipping Business Management	4	62	73	162	39
港口物流设备与自动控制 Harbour Logistics Equipment and Automatic Control	1	114	0	245	96
轮机工程技术 Engine Engineering	29	2 509	4 000	10 051	3 670
船舶检验 Ship Inspection	2	25	45	66	21
船舶工程技术 Ship Engineering	20	1 418	1 934	5 350	1 937
船舶工程技术 Ship Engineering	1	40	46	94	48
船机制造与维修 Ship Engines Manufacturing and Maintenance	2	1	73	81	8

7-6 续表 continued

专 业 Speciality	专业点数（个） Number of Speciality Agencies (unit)	学生数（人）Students (person)			
		毕业生 Graduated	招 生 Entrants	在校生 Enrollment	毕业班学生 Graduation
船艇动力管理 Ship and Boat Power Equipment Management	1	30	48	79	0
海关管理 Customs Management	1	0	1	1	0
报关与国际货运 Customs Declaration/International Freight Transport	36	1 064	1 149	2 746	1 115
水文与水资源类 Hydrology and Water Resources	1	0	0	56	56
水政水资源管理 Water Resources Management by Water Administration	4	39	27	70	21
水运管理 Water Transport Management	1	0	0	11	0
水文与水资源类新专业 New Specialities in Hydrology and Water Resources	1	3	0	2	0
水产养殖技术 Aquaculture Technology	19	130	198	459	149
港口航道与治河工程 Harbour channel and river harnessing projects	1	23	0	5	5

7-7 全国中等职业教育各海洋专业学生情况
Students from Marine Specialities in the National Secondary Vocational Education

专 业 Speciality	专业点数（个） Number of Speciality Agencies (unit)	学生数（人）Students (person)			
		毕业生 Graduated	招 生 Entrants	在校生 Enrollment	毕业班学生 Graduation
合　计　Total	**509**	**36 807**	**37 027**	**104 479**	**45 027**
海水生态养殖 Seawater Ecological Cultivation	22	592	1 767	3 980	698
农林牧渔类新专业 New Specialities of Agriculture, Forestry, Animal Husbaudry and Fishery	72	4 661	4 594	27 384	15 545
水文与水资源勘测 Hydrological and Water Resources Survey	4	230	181	345	1
风电场机电设备运行与维护 Operation and Maintenance of Electromechanical Equipment in the Wind Power Station	45	1 633	3 350	8 176	2 108
船舶制造与修理 Ships Building and Repair	72	4 100	2 775	8 193	2 798
船舶机械装置安装与维修 Installation and Maintenance of Ships' Mechanical Equipment	19	533	442	1 030	276
船舶驾驶 Ship Piloting	99	13 048	12 147	27 469	11 849
轮机管理 Engines Management	67	9 589	8 671	19 598	9 072

7-7 续表 continued

专 业 Speciality	专业点数（个） Number of Speciality Agencies (unit)	学生数（人）Students (person)			
		毕业生 Graduated	招 生 Entrants	在校生 Enrollment	毕业班学生 Graduation
船舶水手与机工 Ship Sailors and Mechanics	39	729	1 203	2 526	963
船舶电气技术 Ship Electric Technology	22	425	325	905	257
船舶通信与导航 Ship Communication and Navigation	1	0	7	7	0
外轮理货 Foreign Ships Freight Forwarding	14	460	265	1 347	574
船舶检验 Ships Test	5	93	194	303	80
港口机械运行与维护 Operation and Maintenance of Harbour Machinery	16	500	872	2 383	569
工程潜水 Engineering Diving	1	18	22	40	18
矿山机电 Mining mechanical electrical equipment	11	196	212	793	219

注：此表不包含技工学校相关数据。
Note: Data related to Technical Schools are not included in the table.

7-8 分地区各海洋专业博士研究生情况
Doctoral Students in Marine Specialities by Regions

地 区 Region	专业点数（个） Number of Speciality Agencies (unit)	学生数（人）Students (person)			
		毕业生 Graduated	招 生 Entrants	在校生 Enrollment	毕业班学生 Graduation
合 计 Total	**125**	**601**	**819**	**3 584**	**1 780**
北 京 Beijing	7	15	24	110	52
天 津 Tianjin	3	1	0	11	11
辽 宁 Liaoning	5	40	56	322	49
上 海 Shanghai	17	48	57	266	155
江 苏 Jiangsu	13	47	72	404	246
浙 江 Zhejiang	3	4	11	40	12
福 建 Fujian	6	23	29	121	63
山 东 Shandong	19	207	265	1 084	583
广 东 Guangdong	11	58	70	259	118
其 他 Others	41	158	235	967	491

7-9 分地区各海洋专业硕士研究生情况
Postgraduate Students in Marine Specialities by Regions

地　区 Region	专业点数（个） Number of Speciality Agencies (unit)	学生数（人）Students (person)			
		毕业生 Graduated	招　生 Entrants	在校生 Enrollment	毕业班学生 Graduation
合　计 **Total**	**294**	**3 034**	**3 469**	**10 354**	**3 564**
北　京 Beijing	19	96	105	309	98
天　津 Tianjin	6	25	37	106	33
河　北 Hebei	5	17	17	55	20
辽　宁 Liaoning	21	335	386	1 104	395
上　海 Shanghai	27	407	354	1 105	434
江　苏 Jiangsu	23	318	390	1 179	382
浙　江 Zhejiang	19	139	183	510	166
福　建 Fujian	15	123	197	520	146
山　东 Shandong	35	370	501	1 510	526
广　东 Guangdong	27	236	280	794	266
广　西 Guangxi	2	25	22	69	24
海　南 Hainan	2	22	27	82	28
其　他 Others	93	921	970	3 011	1 046

7-10 分地区普通高等教育各海洋专业本科学生情况
Undergraduates and Students from Colleges for Professional Training in the Marine Specialities of Ordinary Higher Education by Regions

地 区 Region	专业点数（个） Number of Speciality Agencies (unit)	学生数（人）Students (person)			
		毕业生 Graduated	招 生 Entrants	在校生 Enrollment	毕业班学生 Graduation
合 计 Total	**201**	**12 543**	**15 257**	**58 646**	**14 242**
北 京 Beijing	3	100	158	625	157
天 津 Tianjin	12	592	871	3 206	704
河 北 Hebei	7	110	118	788	215
辽 宁 Liaoning	19	1 960	2 044	8 407	2 225
上 海 Shanghai	15	1 025	1 423	5 207	1 226
江 苏 Jiangsu	23	1 375	1 348	5 668	1 558
浙 江 Zhejiang	20	713	789	2 935	743
福 建 Fujian	8	844	1 215	4 171	1 043
山 东 Shandong	25	1 476	1 714	6 832	1 602
广 东 Guangdong	12	597	898	3 461	782
广 西 Guangxi	5	92	196	659	140
海 南 Hainan	2	115	124	458	102
其 他 Others	50	3 544	4 359	16 229	3 745

7-11 分地区普通高等教育各海洋专业专科学生情况
Undergraduates and Students from Colleges for Professional Training in the Marine Specialities of Ordinary Higher Education by Regions

地 区 Region	专业点数（个） Number of Speciality Agencies (unit)	学生数（人）Students (person)			
		毕业生 Graduated	招 生 Entrants	在校生 Enrollment	毕业班学生 Graduation
合 计 Total	**448**	**37 515**	**33 845**	**106 581**	**36 608**
天 津 Tianjin	14	2 071	1 507	4 859	1 645
河 北 Hebei	23	1 457	802	3 705	1 643
辽 宁 Liaoning	27	1 895	2 974	7 634	2 338
上 海 Shanghai	33	3 098	2 290	7 774	2 811
江 苏 Jiangsu	54	5 748	4 256	15 431	5 570
浙 江 Zhejiang	28	1 984	1 942	5 905	1 917
福 建 Fujian	31	2 059	1 817	5 914	2 146
山 东 Shandong	66	6 645	5 921	18 055	5 854
广 东 Guangdong	28	1 552	2 011	5 569	1 721
广 西 Guangxi	22	938	1 249	3 976	1 431
海 南 Hainan	8	751	358	1 496	614
其 他 Others	114	9 317	8 718	26 263	8 918

7-12 分地区成人高等教育各海洋专业本科学生情况
Students from Marine Specialities in the Adult Higher Education by Regions

地　区 Region	专业点数（个） Number of Speciality Agencies (unit)	学生数（人）Students (person)			
		毕业生 Graduated	招生 Entrants	在校生 Enrollment	毕业班学生 Graduation
合　计 Total	**48**	**1 480**	**2 303**	**5 579**	**2 404**
天　津 Tianjin	2	119	78	223	67
河　北 Hebei	1	0	20	65	27
辽　宁 Liaoning	8	186	314	765	471
上　海 Shanghai	2	83	31	243	72
江　苏 Jiangsu	4	496	1 040	2 492	1 188
浙　江 Zhejiang	5	91	141	300	77
福　建 Fujian	1	0	21	50	13
山　东 Shandong	9	195	94	388	126
广　东 Guangdong	4	79	275	302	0
其　他 Others	12	231	289	751	363

7-13 分地区成人高等教育各海洋专业专科学生情况 Students from Marine Specialities in the Adult Higher Education by Regions

地 区 Region	专业点数（个） Number of Speciality Agencies (unit)	学生数（人）Students (person)			
		毕业生 Graduated	招 生 Entrants	在校生 Enrollment	毕业班学生 Graduation
合 计 Total	**130**	**5 489**	**7 596**	**19 575**	**7 241**
天 津 Tianjin	5	228	40	475	207
河 北 Hebei	2	3	11	196	162
辽 宁 Liaoning	12	1 326	1 781	4 761	2 256
上 海 Shanghai	16	414	891	1 830	591
江 苏 Jiangsu	18	817	1 250	2 754	1 029
浙 江 Zhejiang	12	1 164	1 470	3 101	1 116
福 建 Fujian	5	238	249	740	217
山 东 Shandong	14	187	558	1 271	164
广 东 Guangdong	9	303	56	1 297	546
广 西 Guangxi	3	1	2	4	2
海 南 Hainan	1	3	3	10	3
其 他 Others	33	805	1 285	3 136	948

7-14 分地区中等职业教育各海洋专业学生情况
Students from Marine Specialities in the Secondary Vocational Education by Regions

地 区 Region	专业点数（个） Number of Speciality Agencies (unit)	学生数（人）Students (person)			
		毕业生 Graduated	招 生 Entrants	在校生 Enrollment	毕业班学生 Graduation
合 计 Total	**509**	**36 807**	**37 027**	**104 479**	**45 027**
天 津 Tianjin	6	935	571	1 303	694
河 北 Hebei	35	1 682	2 226	7 946	3 695
辽 宁 Liaoning	47	2 849	1 759	4 608	1 610
上 海 Shanghai	12	539	431	1 143	381
江 苏 Jiangsu	70	2 414	2 534	8 148	2 289
浙 江 Zhejiang	27	2 056	1 543	4 199	945
福 建 Fujian	53	3 866	3 256	10 624	5 577
山 东 Shandong	59	10 995	12 050	24 607	10 154
广 东 Guangdong	22	601	1 335	4 041	1 719
广 西 Guangxi	12	741	1 801	2 763	590
海 南 Hainan	3	50	86	243	60
其 他 Others	163	10 079	9 435	34 854	17 313

7-15 分地区开设海洋专业高等学校教职工数
Number of Teaching and Administrative Staff in the Universities and Colleges Offering Marine Specialities by Regions

地 区 Region	机构数（个） Number of Institutions (unit)	教职工数（人） Number of Teaching and Administrative Staff (person)	专任教师数（人） Number of Full-Time Teacher (person)
合 计 Total	**346**	**413 390**	**254 042**
北 京 Beijing	3	5 326	3 158
天 津 Tianjin	12	15 292	9 905
河 北 Hebei	19	17 063	10 064
辽 宁 Liaoning	19	17 575	10 714
上 海 Shanghai	17	21 610	11 078
江 苏 Jiangsu	40	61 389	39 027
浙 江 Zhejiang	19	24 095	13 480
福 建 Fujian	15	11 966	8 116
山 东 Shandong	47	53 235	33 476
广 东 Guangdong	17	24 355	14 754
广 西 Guangxi	15	12 478	8 673
海 南 Hainan	6	5 915	3 604
其 他 Others	117	143 091	87 993

主要统计指标解释

海洋专业 指高等教育和中等职业教育所设的与海洋有关的专业。

Explanatory Notes on Main Statistical Indicators

Marine Speciality refers to the ocean-related speciality in high learning and the professional secondary vocational education.

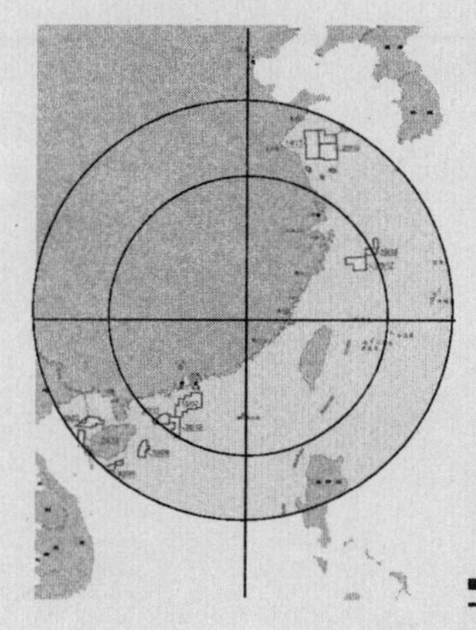

8
海洋环境保护
Marine Environmental Protection

8-1 海区海水水质评价结果
Seawater Quality Assessment Results

项 目 Item	**全国总计 National Total**	渤 海 Bohai Sea	黄 海 Huanghai Sea	东 海 East China Sea	南 海 South China Sea
测站个数 （个） Number of Survey Stations (number)	**7 719**	1 572	1 589	2 496	2 062
清洁海域面积 （万平方千米） Area of Clean Sea Waters (10 000km^2)		4.54			
较清洁海域面积 （万平方千米） Cleaner Sea Area (10 000km^2)	**4.78**	1.47	1.38	1.54	0.39
轻度污染海域面积 （万平方千米） Lightly-Polluted Sea Area (10 000km^2)	**3.43**	0.90	0.72	1.08	0.74
中度污染海域面积 （万平方千米） Moderately-Polluted Sea Area (10 000km^2)	**1.83**	0.38	0.42	0.92	0.12
严重污染海域面积 （万平方千米） Heavily-Polluted Sea Area (10 000km^2)	**4.38**	0.42	0.95	2.73	0.28
首要超标污染物 Prime Pollutants Exceeding the Set Standard	**无机氮、活性磷酸盐、石油类 Inorganic Nitrogen, Active Phosphate, Oils**	无机氮、活性磷酸盐、石油类 Inorganic Nitrogen, Active Phosphate, Oils	无机氮、石油类 Inorganic Nitrogen,Oils	无机氮、活性磷酸盐 Inorganic Nitrogen, Active Phosphate	无机氮、活性磷酸盐、石油类 Inorganic Nitrogen, Active Phosphate, Oils

8-2 海区废弃物海洋倾倒情况
Ocean Dumping of Wastes by Sea Area

海 区 Sea Area	疏浚物 （万立方米） Dredged Materials (10 000m^3)	惰性无机地质废料 （立方米） Inert, Inorganic Geologic Wastes (m^3)
合 计 **Total**	**15 541.23**	**7 311 000**
渤黄海 Bohai and Huanghai Sea	2 929.66	
东 海 East China Sea	9 754.72	
南 海 South China Sea	2 856.85	7 311 000

8-3 海区海洋石油勘探开发污染物排放入海情况
Discharge of Pollutants into the Sea from Offshore Oil Exploration and Exploitation

海 区 Sea Area	生产废水 （万立方米） Sewage from Production (10 000m^3)	泥浆 （立方米） Sludge (m^3)	钻屑 （立方米） Debris from Drilling (m^3)	机舱污水 （立方米） Sewage from Engineroom (m^3)	食品废弃物 （吨） Food Wastes (t)	生活污水 （立方米） Domestic Sewage (m^3)
合 计 Total	**12 858.96**	**47 708.81**	**40 928.76**	**2 226.10**	**1 191.58**	**360 860**
渤 海 Bohai Sea	597.99	7 527.71	26 144.16		100.68	113 545
黄 海 Huanghai Sea						
东 海 East China Sea	135.20	1 291.10	826.40		415.00	32 370
南 海 South China Sea	12 125.77	38 890.00	13 958.20	2 226.10	675.90	214 945

8-4 沿海地区工业废水排放及处理情况
Discharge and Treatment of Industrial Waste Water by Coastal Regions

单位：万吨 (10 000 t)

地 区 Region		工业废水排放总量 Total Volume of Industrial Waste Water Discharged	直排入海 Discharged Directly to Sea
总 计 Total		**1 353 031.21**	**177 516.27**
环渤海经济区 Round-the-Bohai Sea Economic Zone	**合 计 Total**	**416 002.22**	**37 773.25**
	辽 宁 Liaoning	90 457.12	26 734.57
	河 北 Hebei	118 505.34	1 990.44
	天 津 Tianjin	19 795.15	0.54
	山 东 Shandong	187 244.60	9 047.71
长江三角洲经济区 Yangtze River Delta Economic Zone	**合 计 Total**	**473 163.91**	**20 615.49**
	江 苏 Jiangsu	246 298.50	1 115.50
	上 海 Shanghai	44 625.73	10 195.22
	浙 江 Zhejiang	182 239.69	9 304.77
海峡西岸经济区 Economic Zone on West Side of the Straits	**合 计 Total**	**177 185.62**	**107 994.44**
	福 建 Fujian	177 185.62	107 994.44
珠江三角洲经济区 Zhujiang River Delta Economic Zone	**合 计 Total**	**178 625.60**	**6 218.00**
	广 东 Guangdong	178 625.60	6 218.00
环北部湾经济区 Round-the-Beibu Gulf Economic Zone	**合 计 Total**	**108 053.86**	**4 915.10**
	广 西 Guangxi	101 233.74	1 799.51
	海 南 Hainan	6 820.12	3 115.59

8-5 沿海城市工业废水排放及处理情况
Discharge and Treatment of Industrial Waste Water by Coastal Cities

单位：万吨 (10 000 t)

沿海城市 Coastal City	工业废水排放总量 Total Volume of Industrial Waste Water Discharged	
		直排入海 Discharged Directly to Sea
天　津 Tianjin	19 795.15	0.54
唐　山 Tangshan	17 307.77	1 130.00
秦皇岛 Qinhuangdao	6 380.49	246.64
沧　州 Cangzhou	11 365.73	285.89
大　连 Dalian	31 487.21	25 100.94
丹　东 Dandong	5 516.89	18.80
锦　州 Jinzhou	5 497.06	64.90
营　口 Yingkou	3 796.32	955.43
盘　锦 Panjin	3 759.41	0.00
葫芦岛 Huludao	2 769.53	594.48
上　海 Shanghai	44 625.73	10 195.22
南　通 Nantong	19 649.39	133.36
连云港 Lianyungang	6 548.33	497.64
盐　城 Yancheng	21 528.82	475.50

8-5 续表1 continued

沿海城市 Coastal City	工业废水排放总量 Total Volume of Industrial Waste Water Discharged	
		直排入海 Discharged Directly to Sea
杭　州 Hangzhou	47 890.00	710.00
宁　波 Ningbo	19 797.08	6 266.83
温　州 Wenzhou	8 038.29	33.09
嘉　兴 Jiaxing	22 381.10	441.41
绍　兴 Shaoxing	33 123.76	0.00
舟　山 Zhoushan	2 125.07	1 314.73
台　州 Taizhou	6 304.75	538.71
福　州 Fuzhou	5 491.91	362.70
厦　门 Xiamen	31 542.47	26 739.06
莆　田 Putian	2 122.31	253.72
泉　州 Quanzhou	20 715.74	2 643.18
漳　州 Zhangzhou	86 621.76	77 906.12
宁　德 Ningde	1 523.13	72.44
青　岛 Qingdao	11 288.85	1 755.92
东　营 Dongying	9 701.09	0.00
烟　台 Yantai	8 875.37	2 605.23
潍　坊 Weifang	28 191.05	279.82
威　海 Weihai	2 309.18	87.14
日　照 Rizhao	8 361.11	4 022.56
滨　州 Binzhou	14 181.71	0.00

8-5 续表2 continued

沿海城市 Coastal City	工业废水排放总量 Total Volume of Industrial Waste Water Discharged	直排入海 Discharged Directly to Sea
广　州　Guangzhou	24 578.88	62.34
深　圳　Shenzhen	11 715.41	295.16
珠　海　Zhuhai	4 876.14	57.07
汕　头　Shantou	5 788.69	31.82
江　门　Jiangmen	15 590.98	68.99
湛　江　Zhanjiang	6 130.83	724.06
茂　名　Maoming	5 245.24	958.67
惠　州　Huizhou	6 784.70	843.18
汕　尾　Shanwei	1 205.91	7.92
阳　江　Yangjiang	2 154.75	1.50
东　莞　Dongguan	29 103.36	2 679.54
中　山　Zhongshan	9 454.61	0.00
潮　州　Chaozhou	2 890.03	67.37
揭　阳　Jieyang	2 936.28	0.00
北　海　Beihai	1 748.98	25.20
防城港　Fangchenggang	2 909.45	719.52
钦　州　Qinzhou	2 482.32	1 053.50
海　口　Haikou	619.87	0.00
三　亚　Sanya	75.91	0.00

8-6 沿海地带工业废水排放及处理情况
Discharge and Treatment of Industrial Waste Water by Coastal Counties

单位：万吨 (10 000 t)

地 区 Region		工业废水排放总量 Total Volume of Industrial Waste Water Discharged	直排入海 Discharged Directly to Sea
天 津	**Tianjin**	**9 880.82**	**0.54**
河 北	**Hebei**	**10 624.49**	**1 296.64**
唐 山	Tangshan	4 740.67	1 050.00
秦皇岛	Qinhuangdao	4 476.63	246.64
沧 州	Cangzhou	1 407.19	0.00
辽 宁	**Liaoning**	**37 032.09**	**23 563.46**
大 连	Dalian	27 048.36	21 999.55
丹 东	Dandong	1 585.06	18.80
锦 州	Jinzhou	2 639.56	0.00
营 口	Yingkou	2 106.20	955.43
盘 锦	Panjin	1 031.12	0.00
葫芦岛	Huludao	2 621.79	589.68
上 海	**Shanghai**	**30 628.07**	**10 195.22**
江 苏	**Jiangsu**	**27 084.04**	**1 097.22**
南 通	Nantong	9 246.32	124.08
连云港	Lianyungang	4 587.43	497.64
盐 城	Yancheng	13 250.29	475.50

8-6 续表1 continued

地 区 Region	工业废水排放总量 Total Volume of Industrial Waste Water Discharged	直排入海 Discharged Directly to Sea
浙 江 Zhejiang	**77 343.44**	**9 268.27**
杭 州 Hangzhou	14 369.19	710.00
宁 波 Ningbo	17 392.09	6 231.23
温 州 Wenzhou	5 723.43	33.09
嘉 兴 Jiaxing	8 521.83	440.51
绍 兴 Shaoxing	24 643.12	0.00
舟 山 Zhoushan	2 125.07	1 314.73
台 州 Taizhou	4 568.71	538.71
福 建 Fujian	**138 308.57**	**107 977.22**
福 州 Fuzhou	3 940.98	362.70
厦 门 Xiamen	31 542.47	26 739.06
莆 田 Putian	2 122.31	253.72
泉 州 Quanzhou	18 145.66	2 643.18
漳 州 Zhangzhou	81 557.01	77 906.12
宁 德 Ningde	1 000.14	72.44
山 东 Shandong	**44 209.95**	**4 794.29**
青 岛 Qingdao	9 884.71	1 755.92
东 营 Dongying	9 358.09	0.00
烟 台 Yantai	7 329.13	2 605.23
潍 坊 Weifang	10 816.28	0.00
威 海 Weihai	1 782.45	85.65
日 照 Rizhao	1 169.22	347.49
滨 州 Binzhou	3 870.07	0.00

8-6 续表2 continued

地区 Region	工业废水排放总量 Total Volume of Industrial Waste Water Discharged	直排入海 Discharged Directly to Sea
广　东 Guangdong	**95 327.23**	**2 226.59**
广　州 Guangzhou	13 179.38	62.34
深　圳 Shenzhen	10 177.51	295.16
珠　海 Zhuhai	2 640.46	31.15
汕　头 Shantou	5 759.22	31.82
江　门 Jiangmen	10 025.52	68.99
湛　江 Zhanjiang	5 918.19	701.67
茂　名 Maoming	3 363.63	958.67
惠　州 Huizhou	1 456.02	0.00
汕　尾 Shanwei	1 183.27	7.92
阳　江 Yangjiang	944.72	1.50
东　莞 Dongguan	29 103.36	0.00
中　山 Zhongshan	9 454.61	0.00
潮　州 Chaozhou	1 085.45	67.37
揭　阳 Jieyang	1 035.89	0.00
广　西 Guangxi	**5 051.14**	**1 798.22**
北　海 Beihai	1 748.98	25.20
防城港 Fangchenggang	1 886.35	719.52
钦　州 Qinzhou	1 415.81	1 053.50
海　南 Hainan	**652.37**	**351.67**
海　口 Haikou	576.46	0.00
三　亚 Sanya	75.91	351.67

注：1.表中数据为沿海地带合计数（表8-7、8-10、8-13同）。
　　2.沿海地带是指有海岸线的县、县级市、区（包括直辖市和地级市的区）。

Note: 1. The data in the table are the total numbers for the coastal regions (The same for Tables 8-7, 8-10, 8-13).
2.Coastal Zone refers to the counties, county-level cities and districts with coastlines (including the districts under the municipalities directly under the Central Government and the prefecture-level districts)

8-7 沿海地区一般工业固体废物倾倒丢弃、处理及综合利用情况
Discharge, Treatment and Multipurposed Utilization of Common Industrial Solid Wastes by Coastal Regions

单位：吨 (t)

地区 Region		一般工业固体废物倾倒丢弃量 Volume of Common Industrial Solid Wastes Discharged	一般工业固体废物处置量 Volume of Common Industrial Solid Wastes Treated	一般工业固体废物综合利用量 Volume of Common Industrial Solid Wastes Muti-Utilized
总 计 Total		**164 034**	**263 813 763**	**787 005 425**
环渤海经济区 Round-the-Bohai Sea Economic Zone	**合 计 Total**	**85 842**	**213 161 679**	**496 159 838**
	辽 宁 Liaoning	81 781	133 942 910	107 477 799
	河 北 Hebei	4 044	68 064 360	188 211 474
	天 津 Tianjin	0	91 472	17 485 673
	山 东 Shandong	17	11 062 937	182 984 892
长江三角洲经济区 Yangtze River Delta Economic Zone	**合 计 Total**	**9 135**	**7 190 074**	**164 471 026**
	江 苏 Jiangsu	2	3 345 556	99 972 394
	上 海 Shanghai	4 712	748 924	23 581 103
	浙 江 Zhejiang	4 421	3 095 594	40 917 529
海峡西岸经济区 Economic Zone on West Side of the Straits	**合 计 Total**	**8 694**	**13 041 441**	**30 243 259**
	福 建 Fujian	8 694	13 041 441	30 243 259
珠江三角洲经济区 Zhujiang River Delta Economic Zone	**合 计 Total**	**34 088**	**8 099 440**	**51 194 691**
	广 东 Guangdong	34 088	8 099 440	51 194 691
环北部湾经济区 Round-the-Beibu Gulf Economic Zone	**合 计 Total**	**26 275**	**22 321 130**	**44 936 612**
	广 西 Guangxi	25 739	20 504 454	42 923 626
	海 南 Hainan	536	1 816 676	2 012 986

8-8 沿海城市一般工业固体废物倾倒丢弃、处理及综合利用情况
Discharge, Treatment and Multipurposed Utilization of Common Industrial Solid Wastes by Coastal Cities

单位：吨 (t)

沿海城市 Coastal City	一般工业固体废物倾倒丢弃量 Volume of Common Industrial Solid Wastes Discharged	一般工业固体废物处置量 Volume of Common Industrial Solid Wastes Treated	一般工业固体废物综合利用量 Volume of Common Industrial Solid Wastes Muti-Utilized
天　津　Tianjin	0	91 472	17 485 673
唐　山　Tangshan	3 400	28 552 769	71 754 483
秦皇岛　Qinhuangdao	0	6 631 217	11 504 938
沧　州　Cangzhou	0	13 667	3 873 810
大　连　Dalian	0	206 005	5 080 947
丹　东　Dandong	0	4 329 442	1 701 929
锦　州　Jinzhou	0	997 164	2 326 602
营　口　Yingkou	0	11 084	7 605 316
盘　锦　Panjin	0	121 610	1 105 153
葫芦岛　Huludao	91	83 056	3 864 086
上　海　Shanghai	4 712	748 924	23 581 103
南　通　Nantong	0	125 298	4 532 333
连云港　Lianyungang	0	82 757	4 097 589
盐　城　Yancheng	0	42 006	1 709 522

8-8 续表1 continued

沿海城市 Coastal City	一般工业固体废物倾倒丢弃量 Volume of Common Industrial Solid Wastes Discharged	一般工业固体废物处置量 Volume of Common Industrial Solid Wastes Treated	一般工业固体废物综合利用量 Volume of Common Industrial Solid Wastes Muti-Utilized
杭　州 Hangzhou	0	556 703	7 075 762
宁　波 Ningbo	125	1 124 688	12 814 444
温　州 Wenzhou	1 021	158 073	2 268 743
嘉　兴 Jiaxing	0	380 228	3 898 407
绍　兴 Shaoxing	0	305 840	3 211 461
舟　山 Zhoushan	0	49 613	730 335
台　州 Taizhou	2 709	261 289	2 358 820
福　州 Fuzhou	2 703	688 707	6 230 223
厦　门 Xiamen	0	53 756	1 116 733
莆　田 Putian	2 329	5 022	1 568 207
泉　州 Quanzhou	797	250 718	6 722 432
漳　州 Zhangzhou	0	83 343	2 258 664
宁　德 Ningde	61	134 145	790 430
青　岛 Qingdao	0	97 441	8 819 192
东　营 Dongying	0	221 217	2 924 740
烟　台 Yantai	0	3 686 569	19 463 299
潍　坊 Weifang	0	100	8 495 759
威　海 Weihai	0	230 893	2 996 049
日　照 Rizhao	0	44 233	9 522 184
滨　州 Binzhou	0	1 984 448	5 368 127

8-8 续表2 continued

沿海城市 Coastal City	一般工业固体废物倾倒丢弃量 Volume of Common Industrial Solid Wastes Discharged	一般工业固体废物处置量 Volume of Common Industrial Solid Wastes Treated	一般工业固体废物综合利用量 Volume of Common Industrial Solid Wastes Muti-Utilized
广　州 Guangzhou	1	298 356	6 255 131
深　圳 Shenzhen	500	36 118	921 299
珠　海 Zhuhai	0	77 933	3 004 379
汕　头 Shantou	2	5 757	1 096 734
江　门 Jiangmen	42	250 167	2 316 869
湛　江 Zhanjiang	849	93 204	2 492 354
茂　名 Maoming	25	498 803	1 711 930
惠　州 Huizhou	0	53 994	591 175
汕　尾 Shanwei	32	2 930	441 572
阳　江 Yangjiang	2 185	787	1 789 012
东　莞 Dongguan	8 719	1 218 121	4 575 184
中　山 Zhongshan	14 095	60 787	869 657
潮　州 Chaozhou	491	1 605	1 427 502
揭　阳 Jieyang	0	490	925 381
北　海 Beihai	205	2 054 789	1 637 648
防城港 Fangchenggang	0	0	1 713 164
钦　州 Qinzhou	234	22 607	1 347 919
海　口 Haikou	0	4 674	43 018
三　亚 Sanya	0	0	17 000

8-9 沿海地带一般工业固体废物倾倒丢弃、处理及综合利用情况
Discharge, Treatment and Multipurposed Utilization of Common Industrial Solid Wastes by Coastal Counties

单位：吨 (t)

地　区 Region	一般工业固体废物倾倒丢弃量 Volume of Common Industrial Solid Wastes Discharged	一般工业固体废物处置量 Volume of Common Industrial Solid Wastes Treated	一般工业固体废物综合利用量 Volume of Common Industrial Solid Wastes Muti-Utilized
天　津　Tianjin	**5 101 006**	**57 241**	**5 098 740**
	0	0	0
河　北　Hebei	**0**	**1 815 225**	**17 523 235**
唐　山　Tangshan	0	901 288	11 969 244
秦皇岛　Qinhuangdao	0	913 937	5 521 577
沧　州　Cangzhou	0	0	32 414
	0	0	0
辽　宁　Liaoning	**17**	**299 465**	**14 188 209**
大　连　Dalian	0	173 942	4 017 212
丹　东　Dandong	0	61	404 196
锦　州　Jinzhou	0	0	155 712
营　口　Yingkou	0	9 864	7 258 640
盘　锦　Panjin	0	34 103	94 069
葫芦岛　Huludao	17	81 495	2 258 380
上　海　Shanghai	**21 479 713**	**436 720**	**20 954 315**
江　苏　Jiangsu	**0**	**93 413**	**6 156 970**
南　通　Nantong	0	69 464	2 197 800
连云港　Lianyungang	0	8 500	3 024 163
盐　城　Yancheng	0	15 450	935 007

8-9 续表1 continued

地 区 Region	一般工业固体废物倾倒丢弃量 Volume of Common Industrial Solid Wastes Discharged	一般工业固体废物处置量 Volume of Common Industrial Solid Wastes Treated	一般工业固体废物综合利用量 Volume of Common Industrial Solid Wastes Muti-Utilized
浙 江 Zhejiang	**140**	**1 858 370**	**23 331 624**
杭 州 Hangzhou	0	178 536	1 680 918
宁 波 Ningbo	115	1 077 856	12 539 098
温 州 Wenzhou	0	140 012	1 934 622
嘉 兴 Jiaxing	0	93 440	2 543 348
绍 兴 Shaoxing	0	94 977	1 680 790
舟 山 Zhoushan	0	49 613	730 335
台 州 Taizhou	25	223 936	2 222 513
福 建 Fujian	**2 856**	**859 183**	**13 381 231**
福 州 Fuzhou	28	579 184	5 388 722
厦 门 Xiamen	0	53 756	1 116 733
莆 田 Putian	2 329	5 022	1 568 207
泉 州 Quanzhou	499	144 029	3 016 249
漳 州 Zhangzhou	0	76 305	1 589 140
宁 德 Ningde	0	887	702 180
山 东 Shandong	**0**	**4 246 164**	**41 693 459**
青 岛 Qingdao	0	82 702	6 756 319
东 营 Dongying	0	193 782	2 572 686
烟 台 Yantai	0	3 680 531	18 933 419
潍 坊 Weifang	0	0	3 011 681
威 海 Weihai	0	226 915	1 777 702
日 照 Rizhao	0	240	7 330 034
滨 州 Binzhou	0	61 994	1 311 618

8-9 续表2 continued

地 区 Region	一般工业固体废物倾倒丢弃量 Volume of Common Industrial Solid Wastes Discharged	一般工业固体废物处置量 Volume of Common Industrial Solid Wastes Treated	一般工业固体废物综合利用量 Volume of Common Industrial Solid Wastes Muti-Utilized
广 东 Guangdong	**25 842**	**2 394 682**	**23 061 033**
广 州 Guangzhou	1	232 289	5 706 708
深 圳 Shenzhen	500	34 693	901 241
珠 海 Zhuhai	0	31 277	214 663
汕 头 Shantou	2	5 757	1 096 734
江 门 Jiangmen	5	203 755	2 081 970
湛 江 Zhanjiang	328	90 584	2 460 888
茂 名 Maoming	25	498 797	1 616 312
惠 州 Huizhou	0	14 116	157 025
汕 尾 Shanwei	12	2 930	439 310
阳 江 Yangjiang	2 155	322	707 692
东 莞 Dongguan	8 719	1 218 121	4 575 184
中 山 Zhongshan	14 095	60 787	869 657
潮 州 Chaozhou	0	864	1 393 597
揭 阳 Jieyang	0	390	840 052
广 西 Guangxi	**205**	**2 067 803**	**3 574 694**
北 海 Beihai	205	2 054 789	1 637 648
防城港 Fangchenggang	0	0	969 395
钦 州 Qinzhou	0	13 014	967 651
海 南 Hainan	**0**	**4 674**	**43 906**
海 口 Haikou	0	4 674	26 906
三 亚 Sanya	0	0	17 000

8-10 沿海城市工业废气排放及处理情况
Emission and Treatment of Industrial Waste Gas in Coastal Cities

单位：吨 (t)

城市 City	工业废气排放量（万立方米）Industrial Waste Gas Emission (10 000 cu.m)	工业二氧化硫排放量 Industrial Sulphur Dioxide Emission	工业氮氧化物排放量 Industrial Nitrogen Oxides Emission	工业烟（粉）尘排放量 Industrial Soot (Dust) Emission	生活二氧化硫排放量 Household Sulphur Dioxide Emission	生活氮氧化物排放量 Household Nitrogen Oxides Emission	生活烟尘排放量 Household Soot Emission
天津 Tianjin	8 919	221 897	300 404	65 333	8 959	4 447	4 071
秦皇岛 Qinhuangdao	3 201	75 555	59 629	69 760	2 370	708	1 969
大连 Dalian	2 871	132 003	111 836	51 295	14 298	2 403	9 612
上海 Shanghai	13 704	210 092	326 949	66 446	29 860	9 079	14 850
连云港 Lianyungang	831	45 390	34 087	19 133	6 945	904	775
宁波 Ningbo	5 910	152 601	252 583	31 816	2 343	648	461
温州 Wenzhou	1 786	39 982	48 938	24 976	598	400	466
福州 Fuzhou	3 593	91 320	101 834	33 813	1 226	236	547
厦门 Xiamen	1 104	19 108	18 303	3 140	486	54	54
青岛 Qingdao	2 438	75 777	78 225	26 849	27 059	2 601	9 795
烟台 Yantai	2 659	88 186	82 587	33 790	13 073	2 354	9 693
深圳 Shenzhen	1 885	10 435	35 425	1 236	488	285	82
珠海 Zhuhai	1 518	31 784	54 467	10 014	280	118	13
汕头 Shantou	984	26 868	31 595	4 377	455	117	55
湛江 Zhanjiang	953	29 778	28 610	17 421	1 999	363	208
北海 Beihai	501	10 827	18 203	6 170	1 240	141	122
海口 Haikou	28	2 002	438	722	21	60	2

8-11 沿海地区污染治理项目情况
Pollution Treatment Projects in Coastal Regions

单位：个 (number)

地 区 Region		当年安排施工项目 Arranged for Construction in the Year		当年竣工项目 Completed in the Year	
		治理废水 Treatment of Waste Water	治理固体废物 Treatment of Solid Wastes	治理废水 Treatment of Waste Water	治理固体废物 Treatment of Solid Wastes
总 计 Total		**2 326**	**274**	**1 705**	**207**
环渤海经济区 Round-the-Bohai Sea Economic Zone	**合 计 Total**	**584**	**78**	**421**	**56**
	辽 宁 Liaoning	76	11	42	5
	河 北 Hebei	105	20	86	19
	天 津 Tianjin	42	4	35	3
	山 东 Shandong	361	43	258	29
长江三角洲经济区 Yangtze River Delta Economic Zone	**合 计 Total**	**948**	**99**	**744**	**87**
	江 苏 Jiangsu	419	67	353	62
	上 海 Shanghai	40	6	29	4
	浙 江 Zhejiang	489	26	362	21
海峡西岸经济区 Economic Zone on West Side of the Straits	**合 计 Total**	**209**	**27**	**127**	**13**
	福 建 Fujian	209	27	127	13
珠江三角洲经济区 Zhujiang River Delta Economic Zone	**合 计 Total**	**409**	**28**	**287**	**22**
	广 东 Guangdong	409	28	287	22
环北部湾经济区 Round-the-Beibu Gulf Economic Zone	**合 计 Total**	**176**	**42**	**126**	**29**
	广 西 Guangxi	146	37	105	25
	海 南 Hainan	30	5	21	4

8-12 沿海城市污染治理项目情况
Pollution Treatment Projects in Coastal Cities

单位：个 (number)

沿海城市 Coastal City	当年安排施工项目 Arranged for Construction in the Year		当年竣工项目 Completed in the Year	
	治理废水 Treatment of Waste Water	治理固体废物 Treatment of Solid Wastes	治理废水 Treament of Waste Water	治理固体废物 Treatment of Solid Wastes
天　津 Tianjin	42	4	35	3
唐　山 Tangshan	21	4	14	4
秦皇岛 Qinhuangdao	2	1	2	0
沧　州 Cangzhou	3	1	3	1
大　连 Dalian	25	0	9	0
丹　东 Dandong	10	0	5	0
锦　州 Jinzhou	4	2	2	1
营　口 Yingkou	2	0	0	0
盘　锦 Panjin	0	6	0	2
葫芦岛 Huludao	6	1	6	1
上　海 Shanghai	40	6	29	4
南　通 Nantong	84	14	77	13
连云港 Lianyungang	17	6	12	5
盐　城 Yancheng	0	0	0	0

8-12 续表1 continued

沿海城市 Coastal City	当年安排施工项目 Arranged for Construction in the Year		当年竣工项目 Completed in the Year	
	治理废水 Treatment of Waste Water	治理固体废物 Treatment of Solid Wastes	治理废水 Treament of Waste Water	治理固体废物 Treatment of Solid Wastes
杭　州 Hangzhou	58	1	47	0
宁　波 Ningbo	17	0	7	0
温　州 Wenzhou	36	0	29	0
嘉　兴 Jiaxing	40	1	32	1
绍　兴 Shaoxing	99	7	58	7
舟　山 Zhoushan	11	3	5	1
台　州 Taizhou	84	4	81	4
福　州 Fuzhou	9	1	9	1
厦　门 Xiamen	40	4	26	4
莆　田 Putian	2	0	2	0
泉　州 Quanzhou	51	12	15	6
漳　州 Zhangzhou	23	1	21	1
宁　德 Ningde	6	1	5	0
青　岛 Qingdao	6	0	6	0
东　营 Dongying	19	7	15	7
烟　台 Yantai	31	6	25	5
潍　坊 Weifang	56	4	43	2
威　海 Weihai	5	1	4	1
日　照 Rizhao	4	0	4	0
滨　州 Binzhou	41	7	14	1

8-12 续表2 continued

沿海城市 Coastal City	当年安排施工项目 Arranged for Construction in the Year		当年竣工项目 Completed in the Year	
	治理废水 Treatment of Waste Water	治理固体废物 Treatment of Solid Wastes	治理废水 Treament of Waste Water	治理固体废物 Treatment of Solid Wastes
广　州 Guangzhou	20	3	14	2
深　圳 Shenzhen	17	2	10	1
珠　海 Zhuhai	20	1	14	1
汕　头 Shantou	17	1	6	1
江　门 Jiangmen	16	1	12	1
湛　江 Zhanjiang	25	1	18	1
茂　名 Maoming	0	0	0	0
惠　州 Huizhou	12	1	11	1
汕　尾 Shanwei	1	0	1	0
阳　江 Yangjiang	4	0	4	0
东　莞 Dongguan	68	3	54	2
中　山 Zhongshan	0	0	0	0
潮　州 Chaozhou	20	0	8	0
揭　阳 Jieyang	11	0	10	0
北　海 Beihai	1	0	1	0
防城港 Fangchenggang	1	0	0	0
钦　州 Qinzhou	19	9	17	7
海　口 Haikou	9	0	8	0
三　亚 Sanya	0	0	0	0

8-13 沿海地带污染治理项目情况
Pollution Treatment Projects in Coastal Counties

单位：个 (number)

地　区 Region	当年安排施工项目 Arranged for Construction in the Year		当年竣工项目 Completed in the Year	
	治理废水 Treatment of Waste Water	治理固体废物 Treatment of Solid Wastes	治理废水 Treament of Waste Water	治理固体废物 Treatment of Solid Wastes
天　津　Tianjin	**17**	**2**	**11**	**2**
河　北　Hebei	**11**	**1**	**9**	**0**
唐　山　Tangshan	9	0	7	0
秦皇岛　Qinhuangdao	2	1	2	0
沧　州　Cangzhou	0	0	0	0
辽　宁　Liaoning	**40**	**1**	**18**	**1**
大　连　Dalian	25	0	9	0
丹　东　Dandong	9	0	4	0
锦　州　Jinzhou	0	0	0	0
营　口　Yingkou	1	0	0	0
盘　锦　Panjin	0	0	0	0
葫芦岛　Huludao	5	1	5	1
上　海　Shanghai	**16**	**1**	**9**	**0**
江　苏　Jiangsu	**83**	**18**	**74**	**16**
南　通　Nantong	68	13	64	12
连云港　Lianyungang	15	5	10	4
盐　城　Yancheng	0	0	0	0

8-13 续表1 continued

地 区 Region	当年安排施工项目 Arranged for Construction in the Year		当年竣工项目 Completed in the Year	
	治理废水 Treatment of Waste Water	治理固体废物 Treatment of Solid Wastes	治理废水 Treament of Waste Water	治理固体废物 Treatment of Solid Wastes
浙 江 Zhejiang	**180**	**10**	**122**	**8**
杭 州 Hangzhou	4	0	4	0
宁 波 Ningbo	16	0	7	0
温 州 Wenzhou	32	0	25	0
嘉 兴 Jiaxing	15	1	9	1
绍 兴 Shaoxing	68	5	38	5
舟 山 Zhoushan	11	3	5	1
台 州 Taizhou	34	1	34	1
福 建 Fujian	**109**	**15**	**59**	**9**
福 州 Fuzhou	3	0	3	0
厦 门 Xiamen	40	4	26	4
莆 田 Putian	2	0	2	0
泉 州 Quanzhou	45	10	11	4
漳 州 Zhangzhou	18	1	16	1
宁 德 Ningde	1	0	1	0
山 东 Shandong	**94**	**21**	**56**	**13**
青 岛 Qingdao	5	0	5	0
东 营 Dongying	19	7	15	7
烟 台 Yantai	31	5	25	4
潍 坊 Weifang	15	2	7	1
威 海 Weihai	5	1	4	1
日 照 Rizhao	0	0	0	0
滨 州 Binzhou	19	6	0	0

8-13 续表2 continued

地 区 Region	当年安排施工项目 Arranged for Construction in the Year		当年竣工项目 Completed in the Year	
	治理废水 Treatment of Waste Water	治理固体废物 Treatment of Solid Wastes	治理废水 Treament of Waste Water	治理固体废物 Treatment of Solid Wastes
广 东 Guangdong	**184**	**10**	**136**	**9**
广 州 Guangzhou	16	2	13	2
深 圳 Shenzhen	13	1	10	1
珠 海 Zhuhai	13	1	7	1
汕 头 Shantou	17	1	6	1
江 门 Jiangmen	10	1	7	1
湛 江 Zhanjiang	24	1	18	1
茂 名 Maoming	0	0	0	0
惠 州 Huizhou	8	0	7	0
汕 尾 Shanwei	1	0	1	0
阳 江 Yangjiang	1	0	1	0
东 莞 Dongguan	68	3	54	2
中 山 Zhongshan	0	0	0	0
潮 州 Chaozhou	4	0	4	0
揭 阳 Jieyang	9	0	8	0
广 西 Guangxi	**7**	**2**	**6**	**1**
北 海 Beihai	1	0	1	0
防城港 Fangchenggang	1	0	0	0
钦 州 Qinzhou	5	2	5	1
海 南 Hainan	**4**	**0**	**3**	**0**
海 口 Haikou	4	0	3	0
三 亚 Sanya	0	0	0	0

8-14 沿海区域海洋类型自然保护区建设情况
Construction of Marine-Type Nature Reserves

地 区 Region	保护区数量（个） Number of Nature Reserves (number)		按保护级别分（个） By Level of Protection (number)		按保护类型分（个） By Type of Protection (number)				保护区面积（平方千米） Area(km^2)
	已建 Already Established	新建 Newly Established	国家级 National	地方级 Provincial	海洋和海岸生态系统 Marine and Coastal Ecosystem	海洋自然历史遗迹 Marine Natural and Historical Relics	海洋生物多样性 Marine Biodiversity	其他 Other	
合 计 Total	**129**		**33**	**96**	**49**	**26**	**51**	**3**	**462 611**
环渤海经济区 Round-the-Bohai Sea Economic Zone	33		12	21	18	4	11		13 797
长江三角洲经济区 Yangtze River Delta Economic Zone	12		5	7	3	1	5	3	2 465
海峡西岸经济区 Economic Zone on West Side of the Straits	12		3	9	5	2	5		1 018
珠江三角洲经济区 Zhujiang River Delta Economic Zone	49		6	43	20		29		419 874
环北部湾经济区 Round-the-Beibu Gulf Economic Zone	23		7	16	3	19	1		25 457

8-15 沿海地区海洋类型自然保护区建设情况
Construction of Marine-Type Nature Reserves

地 区 Region	保护区数量（个） Number of Nature Reserves (number)		按保护级别（个） By Level of Protection (number)		按保护类型分（个） By Type of Protection (number)				保护区面积（平方千米） Area(km²)
	已建 Already Established	新建 Newly Established	国家级 National	地方级 Provincial	海洋和海岸生态系统 Marine and Coastal Ecosystem	海洋自然历史遗迹 Marine Natural and Historical Relics	海洋生物多样性 Marine Biodiversity	其他 Other	
合 计 Total	**129**		**33**	**96**	**49**	**26**	**51**	**3**	**462 611**
天 津 Tianjin	1		1			1			359
河 北 Hebei	5		1	4	4	1			342
辽 宁 Liaoning	14		6	8	9	2	3		9 022
上 海 Shanghai	4		2	2	1			3	941
江 苏 Jiangsu	5		1	4	1	1	3		833
浙 江 Zhejiang	3		2	1	1		2		691
福 建 Fujian	12		3	9	5	2	5		1 018
山 东 Shandong	13		4	9	5		8		4 074
广 东 Guangdong	49		6	43	20		29		419 874
广 西 Guangxi	3		3		3				460
海 南 Hainan	20		4	16		19	1		24 997

8-16 全国海洋生态监控区基本情况
Basic Condition of the Marine Ecological Monitoring Areas Throughout the Country

生态监控区 Ecological Monitoring Area	所在地 Location	面积（平方千米） Area (km^2)	主要生态系统类型 Major Types of Ecosystem	多样性指数* Diversities Indices		
				浮游植物 Phytoplankton	大型浮游动物 Macrozooplankton	底栖生物 Macrobendthos
双台子河口 Shuangtaizi Estuary	辽宁省 Liaoning Province	3 000	河口 Estuary	1.81	0.30	1.56
锦州湾 Jinzhou Bay	辽宁省 Liaoning Province	650	海湾 Bay	2.39	1.04	0.30
滦河口-北戴河 Luanhe Mouth-Beidaihe	河北省 Hebei Province	900	河口 Estuary	1.24		2.07
渤海湾 Bohai Bay	天津市 Tianjin Municipality	3 000	海湾 Bay	2.69	1.85	2.14
莱州湾 Laizhou Bay	山东省 Shandong Province	3 770	海湾 Bay	1.98	1.74	2.95
黄河口 Yellow River Mouth	山东省 Shandong Province	2 600	河口 Estuary	2.28	1.41	2.72
苏北浅滩 North Jiangsu Bank	江苏省 Jiangsu Province	15 400	滩涂湿地 mudflat and Wetland	1.09	2.29	1.91
长江口 Yangtze River Mouth	上海市 Shanghai Municipality	13 668	河口 Estuary	0.95	2.21	1.92
杭州湾 Hangzhou Bay	上海市 浙江省 Shanghai Municipality Zhejiang Province	5 000	海湾 Bay	1.93	1.80	0.45
乐清湾 Yueqing Bay	浙江省 Zhejiang Province	464	海湾 Bay	2.43	2.75	2.49

8-16 续表　continued

生态监控区 Ecological Monitoring Area	所在地 Location	面积（平方千米） Area (km^2)	主要生态系统类型 Major Types of Ecosystem	多样性指数* Diversities Indices		
				浮游植物 Phytoplankton	大型浮游动物 Macrozooplankton	底栖生物 Macrobendthos
闽东沿岸	福建省	5 063	海　湾	2.44	2.75	2.96
Constal East Fujian	Fujian Province		Bay			
大亚湾	广东省	1 200	海　湾	3.04	2.04	2.04
Daya Bay	Guangdong Province		Bay			
珠江口	广东省	3 980	河　口	2.39	2.38	1.85
Zhujiang River Mouth	Guangdong Province		Estuary			
雷州半岛西南沿岸	广东省	1 150	珊瑚礁			
Southwest Coast of Leizhou Peninsula	Guangdong Province		Coral Reef			
广西北海	广西壮族自治区	120	珊瑚礁 红树林 海草床			
Beihai, Guangxi	Guangxi Zhuang Nationality Autonomous Region		Coral Reef Mangroves Seagrass Bed			
北仑河口	广西壮族自治区	150	红树林			
Beilun River Mouth	Guangxi Zhuang Nationality Autonomous		Mangroves			
海南东海岸	海南省	3 750	珊瑚礁 海草床			
East Coast of Hainan	Hainan Province		Coral Reef Seagrass Bed			
西沙珊瑚礁	海南省	400	珊瑚礁			
Xisha Coral Reef	Hainan Province		Coral Reef			

注:*生物多样性指数：是生物种数和种类间个体数量分配均匀性的综合表现，用Shannon-Wiener多样性指数表征。

Note: Biodiversity index: refers to the comprehensive expression of the distributive homogeneity of the number of biological species and the number of individuals between varieties characterized by the Shannon-Wiener biodiversity index.

8-17　沿海地区风暴潮灾害情况
Survey of Storm Surges by Coastal Regions

受灾地区 Disaster Area	受灾人口（万人）Disaster-stricken Population (10 000 persons)	农作物受灾（千公顷）Disaster-Affected crops (1 000 hm^2)	损失船只（艘）Lost Boat (Number)	海水养殖受灾面积（千公顷）Affected Area of Mariculture (1 000 hm^2)
合　计 Total	**234.68**	**106.27**	**2 074**	**41.78**
天　津 Tianjin				
河　北 Hebei				2.13
上　海 Shanghai	100.00		1	1.50
江　苏 Jiangsu				7.80
浙　江 Zhejiang			148	
福　建 Fujian	56.76		21	1.62
山　东 Shandong			321	4.63
广　东 Guangdong	77.92	106.27	303	17.40
广　西 Guangxi				4.44
海　南 Hainan			1 280	2.26

8-17 续表 continued

受灾地区 Disaster Area	损毁海岸工程 （千米） Destroyed Coastal Engineering Architectures(km)	死亡人数* （人） Death Toll (person)	直接经济损失 （亿元） Direct Economic Loss (100 million yuan)
合 计 Total	**58.03**	**0**	**48.81**
天 津 Tianjin	3.00	0	0.10
河 北 Hebei	0.20	0	1.58
上 海 Shanghai	0.14	0	0.12
江 苏 Jiangsu	9.50	0	0.61
浙 江 Zhejiang	13.75	0	1.92
福 建 Fujian	0.05	0	5.32
山 东 Shandong	2.53	0	5.44
广 东 Guangdong	2.15	0	12.63
广 西 Guangxi	22.75	0	1.15
海 南 Hainan	3.96	0	19.94

注：*包括失踪人数。

Notes:*Includes lost people.

8-18 沿海地区赤潮灾害情况
Survey of Red Tide by Coastal Regions

时间 Date	影响区域 Area	最大面积（平方千米） Disaster Area(km^2)
累 计 In the Aggregate		**6 076**
其中：Including:		
5月11日～5月23日 May.11-May.23	辽宁省鸭绿江口—长海县东北海域 Northeast sea area of Yalujiang River-Mouth—Changhai County, Liaoning Province	4 000
5月13日～6月4日 May.13-Jun.4	浙江省温州苍南石坪附近海域 Sea area near Shiping, Cangnan, Wenzhou of Zhejiang Province	200
5月17日 May.17	浙江省舟山东极海域 Dongji Sea area of Zhoushan, Zhejiang Province	100
6月17日 Jun.17	河北省秦皇岛北戴河鸽子窝至抚宁与昌黎分界线附近海域 Sea area adjacent to the area from Geziwo (Doves' Nest) Beidaihe, Qinhuangdao, to the boundary between Funing and Changli, Hebei Province	180
7月26日～8月7日 Jul.26-Aug.7	福建省厦门同安湾顶琼头海域以及鳄鱼屿以南、集美大桥至五缘湾大桥一带海域 Qiongtou sea area at the head of Tong'an Bay, Xiamen and the sea area south of E'yu (Crocodile) Bay and from Jimei Bridge to Wuyuan Bridge, Fujian Province	105
7月26日～7月27日 Jul.26-Jul.27	浙江省象山港西户港口至乌沙山电厂附近海域 Sea off the area from the Xihu Harbour Mouth, Xiangshan Port to the Wushashan Power Plant, Zhejiang Province	160
8月23日～8月25日 Aug.23-Aug.25	浙江省嵊泗东部海域 East sea area of Shengsi, Zhejiang Province	200
9月23日～9月26日 Sept.23-Sept.26	江苏省海州湾连岛东部和北部海域 East and north sea areas of Liandao, Haizhou Bay, Jiangsu Province	200

主要统计指标解释

1. 工业废水排放量 指经过企业厂区所有排放口排到企业外部的工业废水量。包括生产废水、外排的直接冷却水、超标排放的矿井地下水和与工业废水混排的厂区生活污水,不包括外排的间接冷却水(清污不分流的间接冷却水应计算在内)。

2. 直接排入海的工业废水量 指经企业位于海边的排放口，直接排入海中的废水量。直接排入是指废水经过工厂的排污口直接排入海，而未经过城市下水道或其他中间体，也不受其他水体的影响。

3. 工业废水处理量 指报告期内各种水治理设施实际处理的工业废水量,包括处理后外排的和处理后回用的工业废水量，虽经处理但未达到国家或地方排放标准的废水量也应计算在内。计算时，如遇车间和厂排放口均有治理设施，并对同一废水分级处理时，不应重复计算工业废水处理量。

4. 工业废水处理率 指工业废水处理量占需要处理的工业废水量的百分率。其计算公式是:

工业废水处理率 =（工业废水处理量 / 需处理的工业废水量）×100%

式中，需处理工业废水量 =工业废水排放量+工业废水处理回用量-(工业废水排放达标量-工业废水处理排放达标量)

5. 工业废水排放达标量 指各项指标都达到国家或地方排放标准的外排工业废水量,包括经过处理后外排达标的和未经处理外排达标的两部分。国家排放标准见 GB 8978-88。

6. 工业废水处理排放达标量 指经过各种水治理设施处理后达到国家或地方排放标准的废水排放量。

7. 一般工业固体废物处置量 指将固体废物焚烧或者最终置于符合环境保护规定要求的场所并不再回取的工业固体废物量(包括当年处置往年的工业固体废物累计贮存量)。

8. 一般工业固体废物倾倒丢弃量 指将所产生的固体废物倾倒丢弃到固体废物污染防治设施、场所以外的量。不包括矿山开采的剥离废石和掘进废石(煤矸石和呈酸性或碱性的废石除外)。

9. 当年开工污染治理项目数 指报告期内由国家、部门、地方或企业单位安排开工的，并以治理废水、废气、固体废物、噪声和其他(如电磁波、恶臭等)环境污染的环境治理工程的总数。不包括“三同时”项目。

10. 当年竣工项目数 指报告期内竣工投入运行的治理废水、废气、固体废物、噪声及其他污染的环境工程项目的总数。

Explanatory Notes on Main Statistical Indicators

1. Volume of Industrial Waste Water Discharged refers to the quantity of industrial waste water discharged externally through all the outlets in the factory area of the enterprise, including the waste water from production, externally discharged direct cooling water, mine-shaft groundwater discharged exceeding the set standard and the domestic sewage of the factory area discharged together with the industrial waste water, but not including the indirect cooling water discharge externally.

2. Volume of Industrial Waste Water discharged Directly to Sea refers to the quantity of waste water directly discharged into the sea through the outlets of the enterprise by the sea. Direct discharge means the direct discharge into the sea of waste water through the outlets of the factory, which is not discharged via the urban sewers or other intermediates and is not affected by other water bodies.

3. Volume of Industrial Waste Water Treated refers to the industrial waste water volume actually treated by various water treatment facilities in the period covered by the report, including the quantity of the industrial waste water discharged and reused after treatment. The amount of the waste water which is not up to the state or local standard of discharge upon treatment should be included. If the workshops and outlets of the factory are provided with treatment facilities and carry out graded treatment of the same waste water, the processed volume of industrial waste water cannot be calculated repeatedly.

4. Rate of Industrial Waste Water Treatment refers to the percentage of the processed volume of industrial waste water in the industrial waste water volume that needs to be treated. The calculation formula is:

Rate of Industrial Waste Water Treatment = (Processed Volume of Industrial Waste Water/Processed Volume of Industrial Waste Water Required) × 100%.

Where: Processed Volume of Industrial Waste Water Required = (Processed Volume of Industrial Waste Water + Reused Volume of Industrial Waste Water After Treatment−Up-to-Standard Volume of Industrial Waste Water for Discharge−Up-to-Standard Volume of Industrial Waste Water for Discharge after Treatment).

5. Volume of Up-to-Standard Industrial Waste Water Discharged refers to the externally discharged volume of industrial waste water with all its indexes up to the state or local standard for discharge, including the volume of industrial waste water up to the standard for discharge after being treated or that without being treated. See GB8978–88 for the state′s discharge standard.

6. Volume of Up-to-Standard Discharged Industrial Waste Water After Treatment refers to the discharged volume of waste water that is up to the state or local standard for discharge upon treatment by various water treatment facilities.

7. Volume of Common Industrial Solid Wastes Discharged refers to the amount of solid wastes discharged out of the facilities and sites for the solid wastes pollution prevention and control, not including the stripped and tunneled waste ores in the excavation of mines (other than gangues and acid or alkaline waste ores).

8. Volume of Common Industrial Solid Wastes Treated refers to the volume of industrial solid wastes which are to be burned or finally placed at the sites in keeping with the requirement of environmental protection and will not be recovered (including the accumulated amount of storage in former years of industrial solid radioactive matter disposed of in the year).

9. Number of Pollution Treatment Projects Started in the Current Year refers to the total number of environmental pollution control projects for controlling waste water, waste gas, solid wastes, noise and other environmental pollutions (such as electromagnetic wave, offensive odor) started by the state, governmental departments, local governments or enterprises in the period covered by the report mainly for the purpose of pollution control and multipurpose utilization of ″three wastes″.

10. Number of Pollution Treatment Projects Completed in the Current Year refers to the total number of environmental engineering projects for controlling waste water, waste gas, solid wastes, noise and other pollutions which are completed and put into operation in the period covered by the report.

9 海洋行政管理及公益服务

Marine Administration and Public Service

9-1 海域使用管理情况
Sea Area Use Management

地 区 Region	发放海域使用权证书（本） Certificates of Right of Sea Area Use Issued (number)	确权海域面积（公顷） Area of Waters with Established Rights (hm^2)	海域使用金（万元） Charge for Sea Area Utilization (10 000 yuan)
全国总计 National Total	**3 874**	**185 946.16**	**964 493.47**
天 津 Tianjin	33	1 330.67	149 211.22
河 北 Hebei	1 636	24 419.71	109 429.35
辽 宁 Liaoning	346	74 838.72	181 209.97
上 海 Shanghai	4	340.02	1 049.43
江 苏 Jiangsu	282	41 602.05	88 733.66
浙 江 Zhejiang	196	9 839.86	124 026.26
福 建 Fujian	286	5 891.36	47 258.04
山 东 Shandong	227	12 013.87	106 038.96
广 东 Guangdong	333	7 252.03	55 391.70
广 西 Guangxi	434	6 340.28	58 373.24
海 南 Hainan	92	1 561.27	34 397.35
其 他* other	5	516.32	9 374.29

注：*为沿海省（自治区、直辖市）管理海域以外（指渤海中部海域）。

Note: * Sea areas outside the control of coastal provinces of autonomous regions and municipalities directly under the Central Government. (Referring to the mid-Bohai Sea area)

9-2 海洋倾废管理情况
Management on the Ocean Dumping of Wastes

海　区 Sea Area	签发疏浚物海洋倾倒许可证（份） Permit Issued for Ocean Dumping of Dredged Materials (number)	新选划倾倒区数（个） Number of Newly Designated Dumping Zones (number)
合　计 Total	**95**	**6**
渤黄海 Bohai and Huangai Sea	34	2
东　海 East China Sea	16	2
南　海 South China Sea	45	2

9-3 海洋执法检查情况
Marine Law Enforcement Inspection

执法领域 Law Enforcement Area	检查项目（个） Number of Items Subject to Inspection (number)	检查次数（次） Number of Times of Inspection (number)	发现违法行为（起） Illegal Acts Found (case)	作出行政处罚（件） Inflict Administrative Punishments (cases)
总　计 Total	**49 335**	**145 598**	**2 113**	**1 202**
海域使用 Sea Area Use	30 900	81 995	1 433	748
涉外海洋科研 Foreign-related Marine Scientific Research	35	67	0	0
海底电缆保护 Submarine Cable Protection	142	1 245	12	4
海洋工程建设项目 Marine Engineering Construction Projects	5 542	29 915	253	174
海洋倾废 Oceanic Dumping of Wastes	1 419	7 403	155	123
海洋生态保护 Marine Ecological Conservation	618	4 147	193	145
海岛保护 Island Protection	10 679	20 826	67	8

9-4 沿海地区海滨观测台站分布概况
Distribution of Coastal Observation Stations by Coastal Regions

单位：个 (number)

地 区 Region	合 计 Total	海洋站 Marine Station	验潮站① Tide Station	气象台站② Meteorological Station	地震台站 Seismic Station	雷达站 Radar Station
合 计 Total	**868**	**105**	**247**	**91**	**282**	**143**
天 津 Tianjin	**13**	1	0	1	7	4
河 北 Hebei	**33**	4	0	2	21	6
辽 宁 Liaoning	**64**	7	3	14	29	11
上 海 Shanghai	**92**	6	61	1	10	14
江 苏 Jiangsu	**71**	4	26	19	18	4
浙 江 Zhejiang	**99**	21	35	3	21	19
福 建 Fujian	**117**	15	11	23	47	21
山 东 Shandong	**114**	17	21	14	39	23
广 东 Guangdong	**184**	16	82	4	56	26
广 西 Guangxi	**33**	5	5	9	8	6
海 南 Hainan	**47**	9	3	0	26	9

注： ①潮流量观测站43处，潮水位观测站204处。
②中国气象局所属的气象台站仅指沿海气象台站中有海洋气象观测、预报及服务等业务的台站，合计中含有中央气象台1个。

Notes: ①There are 43 tidal current observation stations and 204 tidal level observation stations.
② The meteorological observatories and stations under the China Meteorological Administration only refer to those with marine meteorotogical observation, forecast and service among the coastal meteorological observatories and the total includes the central meteorological station.

9-5 海洋预报服务概况
Marine Forecast Service

单位：次 (time)

预报项目 Item	数值预报 Numerical Forecast			
	预报服务次数 Frequency	发布次数 Frequency of Release		
		广播电视 Radio and TV	因特网 Internet	纸 质 Paper Media
合 计 Total	**12 519**	**256**	**11 460**	**803**
海 浪 Sea Wave	3 285		3 285	
海 温 Sea Surface Temperature	1 124		1 124	
潮 汐 Tide	1 825		1 825	
海 流 Sea Current	3 129	256	2 873	
海平面 Sea Level				
盐 度 Salinity	730		730	
赤 潮 Red Tide	2			2
滨海旅游 Coastal Tourism				
海 冰 Sea Ice	85		85	
绿 潮 Green Seaweed	1 207		932	275
溢 油 Oil Spill	521		510	11
厄尔尼诺 EL Niño				
专 项 Special Item	611		96	515
其 他 Others				

9-5 续表 continued

预报项目 Item	统计预报 Statistical Forecast			
	预报服务次数 Frequency	发布次数 Frequency of Release		
		广播电视 Radio and TV	因特网 Internet	纸 质 Paper Media
合 计 Total	**163 294**	**57 456**	**46 359**	**59 479**
海 浪 Sea Wave	53 358	19 511	19 535	14 312
海 温 Sea Surface Temperature	25 602	16 310		9 292
潮 汐 Tide	37 997	14 361	14 199	9 437
海 流 Sea Current				
海平面 Sea Level				
盐 度 Salinity	36			36
赤 潮 Red Tide	4 382	1 825	470	2 087
滨海旅游 Coastal Tourism	7 746	4 526	2 855	365
海 冰 Sea Ice	3 259	127	244	2 888
绿 潮 Green Seaweed				
溢 油 Oil Spill				
厄尔尼诺 EL Niño	413			413
专 项 Special Item	26 504	796	8 691	17 017
其 他 Others	3 997		365	3 632

注：本表只包含国家海洋局资料。

Note: This table only contains the data from the State Oceanic Administration.

9-6　海洋环境观测情况
Condition of Marine Environmental Observation

项目 Item	台站观测 Station Observation	断面观测 Sectional Observation	浮标观测 Buoy Observation	船舶测报 Ship Measuring and Reporting	其他观测 Other Observation
测站（点）数（个） Number of Station (unit)	105	15	19	57	168
实际获得数据量（个） Quantity of Data Actually Obtained (unit)	135 571 236	38 634	2 664 146	8 814 418	47 489 174

9-7　海洋调查概况
Marine Survey Statistics

调查名称 Name	站点数（个） Number of Stations (unit)	船舶数（艘） Number of Ships (unit)	项目数（个） Number of Items (unit)	实际获得数据（个） Quantity of Data Actually Obtained (unit)	发布通（公、简）报量（期） Quantity of Circulars (Bulletin, Brief Reports) (number)
合　计 Total	**9 049**	**398**	**175**	**20 722 905**	**65**
大洋调查 Oceanic Survey	1 691	2	11	652 906	15
极地调查 Polar Survey	89	2	17	309	
专项调查 Special Survey	6 837	116	61	20 017 695	
其他调查 Other Survey	432	278	86	51 995	50

9-8 涉外海洋科学研究审批情况
Examination and Approval on Foreign-Related Marine Scientific Research

审批单位 Examing and Approving Units	审批研究活动申请（份） Application for the Examing and Approving Research Activity (unit)	船只作业计划审批（份） Examination and Approval of the Ship Operating Plan (unit)
国家海洋局 State Oceanic Administration , People's Republic of China	6	6

9-9 海洋档案及利用情况
Marine File and Its Use

指　标 Item	指 标 值 Data
室（馆）存档案 Files Deposited in the Archives	
纸介质档案（卷、册） Paper Media Files (reel,volume)	141 298
电子档案（GB） Electronic Media Files (GB)	12 834
本年接收档案 Files received this year	
纸质档案（卷、册） Paper Media Files (reel,volume)	8 337
电子档案（GB） Electronic Media Files (GB)	788
本年利用档案 Files utilized this year	
纸质档案（卷、册） Paper Media Files (reel,volume)	11 931
电子档案（GB） Electronic Media Files (GB)	559

9-10 卫星遥感接收应用情况
Remote-Sensing Receiving and Utilization

指　标 Item	指标值 Data
卫星接收次数（轨） Number of Satellite Receptions (orbit / track)	105 762
全年接收时间（天） Whole Year Receiving Time (day)	271 652
全年实际接收存档数据量（GB） Amount of Data Actually Received and Placed on File in the Whole Year (GB)	33 828.24
累计存档数据量（GB） Total Amount of Data Placed on File (GB)	45 294.17
卫星数据分发 Satellite Data Distribution	
类别用户（个） Classified Users (number)	30
分发数据量（GB） Amount of Data Distributed (GB)	22 487.63

9-11 海洋标准化监督管理情况
Supervision and Management of Marine Standardization

单位：项 (item)

指 标 Item	指 标 值 Data
标准立项审查 Examination of the Standards for Authorization	25
国家标准 National Standards	
行业标准 Professional Standards	25
标准审查 Examination of Standards	259
国家标准 National Standards	72
行业标准 Professional Standards	187
标准出版 Standards Publication	12
国家标准 National Standards	1
行业标准 Professional Standards	11
标准实施监督检查（次） Implementation Supervision and Examination of Standards (time)	

主要统计指标解释

1. 海域使用检查 针对不同类型的用海行为进行的监督检查。

2. 涉外海洋科研项目检查 主要针对国际组织、外国组织和个人为和平目的，单独或者与中华人民共和国的组织合作，使用船舶或者其他运载工具、设施，在中华人民共和国内海、领海以及中华人民共和国管辖的其他海域内进行的对海洋环境和海洋资源等的调查研究活动进行监督检查。

3. 海底电缆管道检查 主要是针对铺设海底电缆管道路由调查、铺设施工和维修改造等的监督检查。

4. 海洋工程建设项目环境保护检查 主要是针对防治海洋工程建设项目对海洋环境的污染损害的监督检查。

5. 海洋倾废检查 主要是针对防治倾倒废弃物对海洋环境的污染损害的监督检查。

6. 海洋生态保护检查 主要是针对红树林、珊瑚礁、滨海湿地、海岛、海湾、入海河口、重要渔业水域等具有典型性、代表性的海洋生态系统，珍稀、濒危海洋生物的天然集中分布区，具有重要经济价值的海洋生物生存区域及有重大科学文化价值的海洋自然历史遗迹和自然景观等海洋自然保护区以及其他需要予以特殊保护的区域的监督检查。

7. 断面监测 按照国家海洋局“断面监测方案”，每年定期利用船舶在沿海设定的断面上进行海洋水文、气象、生物、化学等项目的监测活动。

8. 浮标监测 在海上固定站位获取长期、连续海洋环境观测资料的海上锚定资料浮标。

9. 船舶监测 利用在固定航线上走航的商船或渔船，每日定时所在地的海洋环境状况（主要是水文气象要素），并将监测数据实时发送给有关单位，供海洋环境预报使用。

10. 大洋调查 以大洋科考、研究为目的的远洋调查。

11. 专项调查 为完成国家专项任务进行的海洋调查。

12. 全年接收时间 指全年整个应用系统接收的各类遥感卫星数据时，接收设备工作时间。

13. 全年实际接收存档数据量 指全年整个应用系统接收的各类遥感卫星数据原始数据量。

14. 累计存档数据量 指全年整个应用系统制作的各类遥感产品数据量。

15. 分发数据量 指应用系统提供数据产品用于开展各类应用的数据量。

16. 卫星接收次数 地面观测系统接收卫星数据的数量。

17. 国家标准 针对海洋领域内需要在全国范围内统一的有关技术要求所制定的国家标准。海洋国家标准由国家标准化主管部门统一批准、编号和发布。

18. 行业标准 对没有海洋国家标准而又需要在海洋领域内统一的技术要求所制定的标准。海洋行业标准由国家海洋局统一批准、编号和发布。

Explanatory Notes on Main Statistical Indicators

1. Sea Area Use Supervision and inspection refers to the various types of sea area use conducts.
2. Inspection of Foreign-Related Marine Scientific Research Projects refers to the supervision and inspection of the activities of surveying marine environment and resources conducted by international organizations, foreign organizations and individuals for peaceful purposes, alone or in cooperation with PRC organizations, by using ships or other means of delivery as facilities in the internal seas and territorial waters of the People's Republic of China as well as in the other water under the jurisdiction of the People's Republic of China.
3. Inspection of Submarine Cables and Pipelines is mainly aimed at the survey of the submarine cables and pipelines laying route and supervision and inspection of the laying construction, repair and transformation.
4. Environmental Protection Inspection for the Marine Engineering Construction Projects is mainly directed against preventing and controlling the pollution damage of the marine engineering construction project to the marine environment.
5. Inspection of Oceanic Dumping of Wastes mainly refers to the supervision and inspection aimed at preventing and controlling the pollution damage of wastes dumping to the marine environment.
6. Inspection of Marine Ecological Protection mainly refers to the supervision and inspection of the typical and representative marine ecosystems such as mangroves, coral reef, coastal wetland, sea island, bay, estuaries open to the sea, important fishery waters, etc, the natural centralized distribution zones of rare and endangered marine life, the living areas for the marine life with significant economic values and the marine nature reserves such as the marine natural and historical remains and natural landscapes with important scientific and cultural values as well as other areas needing to be specially protected.
7. Sectional Monitoring According to the *"Sectional Monitoring Plan"* of the State Oceanic Administration, monitoring activities concerning such items as marine hydrology, meteorology, biology and chemistry are carried out regularly every year on the sections set in the coastal area.
8. Buoy Monitoring refers to the monitoring carried out by the offshore mooring data buoys which acquire long-term, continuous marine environmental observations at the fixed stations at sea.
9. Ship Monitoring Monitoring the marine environmental condition in the sea area of fixed times every day by using the merchant or fishing vessels cruising on the fixed navigation line (mainly the hydro meteorological elements) and transmitting the monitored data in real time to the related units for use in the marine environmental forecast.
10. Oceanic Survey refers to the oceanic surveys aimed at the oceanic scientific investigations and research.
11. Special-Subject Survey refers to the oceanic investigation for the purpose of fulfilling the state′s

special tasks

12. Whole Year Receiving Time refers to the operating time of receiving equipment in the whole year when the whole application system receives all types of remote sensing satellite data.

13. Amount of Data Actually Received and Placed on File Throughout the year refers to the raw data of remote-sensing satellite data of various kinds received by the whole application system throughout the year.

14. Total Amount of Date Placed on File refers to the date amount of various remote-sensing products made by the whole application systems throughout the year.

15. Amount of Data Distributed refers to the amount of data products provided by the application system for various uses.

16. Number of Satellite Receptions refers to the amount of data received by the ground observation system.

17. National Standards refers to the standards formulated in view of the relevant technical requirements in the marine field that need to be unified throughout the country. The Marine National standards are approved, numbered and issued uniformly by the state department responsible for standardization.

18. Professional Standards refers to the standards formulated for the technical requirements which have no national standards but need to be unified in the marine field. The Marine Professional Standards are approved, numbered and issued by the State Oceanic Administration.

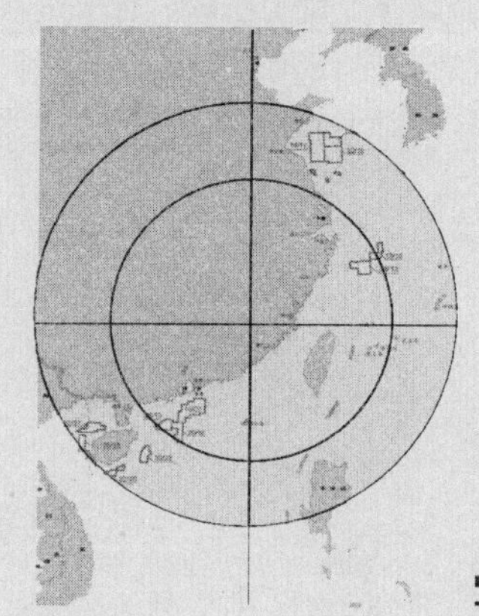

10

全国及沿海社会经济

National and Coastal Socioeconomy

10-1 国内生产总值
Gross Domestic Product

单位：亿元 (100 million yuan)

年 份 Year	国内生产总值 Gross Domestic Product	第一产业 Primary Industry	第二产业 Secondary Industry	第三产业 Tertiary Industry
2001	109 655.2	15 781.3	49 512.3	44 361.6
2002	120 332.7	16 537.0	53 896.8	49 898.9
2003	135 822.8	17 381.7	62 436.3	56 004.7
2004	159 878.3	21 412.7	73 904.3	64 561.3
2005	184 937.4	22 420.0	87 598.1	74 919.3
2006	216 314.4	24 040.0	103 719.5	88 554.9
2007	265 810.3	28 627.0	125 831.4	111 351.9
2008	314 045.4	33 702.0	149 003.4	131 340.0
2009	340 902.8	35 226.0	157 638.8	148 038.0
2010	401 512.8	40 533.6	187 383.2	173 596.0
2011	472 881.6	47 486.2	220 412.8	204 982.5

10-2 国内生产总值增长速度
Growth Rate of Gross Domestic Product

单位：% (%)

年 份 Year	国内生产总值 Gross Domestic Product	第一产业 Primary Industry	第二产业 Secondary Industry	第三产业 Tertiary Industry
2001	8.3	2.8	8.4	10.3
2002	9.1	2.9	9.8	10.4
2003	10.0	2.5	12.7	9.5
2004	10.1	6.3	11.1	10.1
2005	11.3	5.2	12.1	12.2
2006	12.7	5.0	13.4	14.1
2007	14.2	3.7	15.1	16.0
2008	9.6	5.4	9.9	10.4
2009	9.2	4.2	9.9	9.6
2010	10.4	4.3	12.3	9.8
2011	9.3	4.3	10.3	9.4

注：本表按可比价格计算(上年为基期)。
Note:This table is calculated at the comparable price (with the previous year as the base period).

10-3 沿海地区生产总值
Gross Regional Product of Coastal Regions

单位：亿元 (100 million yuan)

地 区 Region	地区生产总值 Gross Regional Product	第一产业 Primary Industry	第二产业 Secondary Industry	第三产业 Tertiary Industry
合 计 Total	**289 050.39**	**20 711.52**	**146 817.59**	**121 521.28**
天 津 Tianjin	11 307.28	159.72	5 928.32	5 219.24
河 北 Hebei	24 515.76	2 905.73	13 126.86	8 483.17
辽 宁 Liaoning	22 226.70	1 915.57	12 152.15	8 158.98
上 海 Shanghai	19 195.69	124.94	7 927.89	11 142.86
江 苏 Jiangsu	49 110.27	3 064.77	25 203.28	20 842.21
浙 江 Zhejiang	32 318.85	1 583.04	16 555.58	14 180.23
福 建 Fujian	17 560.18	1 612.24	9 069.20	6 878.74
山 东 Shandong	45 361.85	3 973.85	24 017.11	17 370.89
广 东 Guangdong	53 210.28	2 665.20	26 447.38	24 097.70
广 西 Guangxi	11 720.87	2 047.23	5 675.32	3 998.33
海 南 Hainan	2 522.66	659.23	714.50	1 148.93

10-4 沿海地区生产总值增长速度
Growth Rate of Gross Regional Product of Coastal Regions

单位：% (%)

地 区 Region	2002	2003	2004	2005	2006	2007	2008	2009	2010	2011
天 津 Tianjin	12.7	14.8	15.8	14.9	14.7	15.5	16.5	16.5	17.4	16.4
河 北 Hebei	9.6	11.6	12.9	13.4	13.4	12.8	10.1	10.0	12.2	11.3
辽 宁 Liaoning	10.2	11.5	12.8	12.7	14.2	15.0	13.4	13.1	14.2	12.2
上 海 Shanghai	11.3	12.3	14.2	11.4	12.7	15.2	9.7	8.2	10.3	8.2
江 苏 Jiangsu	11.7	13.6	14.8	14.5	14.9	14.9	12.7	12.4	12.7	11.0
浙 江 Zhejiang	12.6	14.7	14.5	12.8	13.9	14.7	10.1	8.9	11.9	9.0
福 建 Fujian	10.2	11.5	11.8	11.6	14.8	15.2	13.0	12.3	13.9	12.3
山 东 Shandong	11.7	13.4	15.4	15.0	14.7	14.2	12.0	12.2	12.3	10.9
广 东 Guangdong	12.4	14.8	14.8	14.1	14.8	14.9	10.4	9.7	12.4	10.0
广 西 Guangxi	10.6	10.2	11.8	13.1	13.6	15.1	12.8	13.9	14.2	12.3
海 南 Hainan	9.6	10.6	10.7	10.5	13.2	15.8	10.3	11.7	16.0	12.0

注：本表按可比价格计算(上年为基期)。

Note: This table is calculated at the comparable price (with the previous year as the base period).

10-5 沿海城市生产总值（2010年）
Gross Regional Product of Coastal Cities, 2010

单位：亿元 (100 million yuan)

沿海城市 Coastal City		地区生产总值 Gross Regional Product	第一产业 Primary Industry	第二产业 Secondary Industry	第三产业 Tertiary Industry
合　计	**Total**	**144 031.5**	**7 950.3**	**71 792.7**	**64 288.5**
天　津	**Tianjin**	**9 224.5**	**145.6**	**4 840.2**	**4 238.7**
河　北	**Hebei**	**7 602.8**	**801.3**	**4 081.4**	**2 720.2**
唐　山	Tangshan	4 469.2	421.9	2 598.4	1 448.9
秦皇岛	Qinhuangdao	930.5	126.7	367.8	436.0
沧　州	Cangzhou	2 203.1	252.7	1 115.2	835.3
辽　宁	**Liaoning**	**9 260.0**	**826.8**	**4 850.5**	**3 582.7**
大　连	Dalian	5 158.2	345.1	2 624.5	2 188.6
丹　东	Dandong	728.9	100.1	373.2	255.6
锦　州	Jinzhou	912.6	151.3	434.6	326.8
营　口	Yingkou	1 002.5	77.2	554.7	370.6
盘　锦	Panjin	926.3	81.5	616.5	228.3
葫芦岛	Huludao	531.5	71.6	247.0	212.8
上　海	**Shanghai**	**17 166.0**	**114.2**	**7 218.3**	**9 833.5**
江　苏	**Jiangsu**	**6 991.8**	**823.0**	**3 550.3**	**2 618.5**
南　通	Nantong	3 465.7	266.2	1 908.6	1 290.9
连云港	Lianyungang	1 193.3	182.6	545.1	465.6
盐　城	Yancheng	2 332.8	374.2	1 096.6	862.0
浙　江	**Zhejiang**	**22 203.4**	**1 020.3**	**11 702.1**	**9 481.0**
杭　州	Hangzhou	5 949.2	208.4	2 844.1	2 896.7
宁　波	Ningbo	5 163.0	219.1	2 870.7	2 073.2
温　州	Wenzhou	2 925.1	93.7	1 533.5	1 297.9
嘉　兴	Jiaxing	2 300.2	127.0	1 339.6	833.6
绍　兴	Shaoxing	2 795.2	149.7	1 566.6	1 078.9
舟　山	Zhoushan	644.3	62.0	293.3	289.0
台　州	Taizhou	2 426.4	160.4	1 254.3	1 011.7
福　建	**Fujian**	**11 768.2**	**917.2**	**6 017.6**	**4 833.4**
福　州	Fuzhou	3 123.4	282.7	1 401.9	1 438.8
厦　门	Xiamen	2 060.1	23.1	1 024.5	1 012.5
莆　田	Putian	850.4	87.9	477.1	285.4
泉　州	Quanzhou	3 565.0	132.2	2 144.9	1 287.9
漳　州	Zhangzhou	1 430.7	254.7	652.0	524.0
宁　德	Ningde	738.6	136.6	317.2	284.8

10-5 续表 continued

沿海城市 Coastal City		地区生产总值 Gross Regional Product	第一产业 Primary Industry	第二产业 Secondary Industry	第三产业 Tertiary Industry
山　东	**Shandong**	**19 996.9**	**1 439.1**	**11 253.5**	**7 304.3**
青　岛	Qingdao	5 666.2	277.0	2 758.6	2 630.6
东　营	Dongying	2 360.0	87.4	1 712.2	560.4
烟　台	Yantai	4 358.5	334.5	2 566.5	1 457.5
潍　坊	Weifang	3 090.9	330.5	1 720.3	1 040.1
威　海	Weihai	1 944.6	153.9	1 087.0	703.7
日　照	Rizhao	1 025.2	100.3	561.6	363.3
滨　州	Binzhou	1 551.5	155.5	847.3	548.7
广　东	**Guangdong**	**37 715.9**	**1 511.6**	**17 533.6**	**18 670.6**
广　州	Guangzhou	10 748.4	188.6	4 002.3	6 557.5
深　圳	Shenzhen	9 581.6	6.5	4 523.4	5 051.7
珠　海	Zhuhai	1 208.6	32.4	662.0	514.2
汕　头	Shantou	1 209.0	64.5	678.2	466.2
江　门	Jiangmen	1 570.4	117.0	872.2	581.2
湛　江	Zhanjiang	1 405.1	289.3	577.6	538.2
茂　名	Maoming	1 492.1	274.5	590.8	626.8
惠　州	Huizhou	1 730.0	102.4	1 019.6	608.0
汕　尾	Shanwei	465.1	77.6	212.7	174.8
阳　江	Yangjiang	639.8	140.3	271.6	227.9
东　莞	Dongguan	4 246.5	16.6	2 160.8	2 069.1
中　山	Zhongshan	1 850.6	50.7	1 074.1	725.8
潮　州	Chaozhou	559.2	40.3	309.3	209.6
揭　阳	Jieyang	1 009.5	110.9	579.0	319.6
广　西	**Guangxi**	**1 242.6**	**266.8**	**546.2**	**429.6**
北　海	Beihai	401.5	87.2	167.9	146.4
防城港	Fangchenggang	320.4	47.4	159.8	113.2
钦　州	Qinzhou	520.7	132.2	218.5	170.0
海　南	**Hainan**	**859.4**	**84.4**	**199.0**	**576.0**
海　口	Haikou	617.2	46.0	148.8	422.4
三　亚	Sanya	242.2	38.4	50.2	153.6

注：本表各省数据为合计数。

Note: The data for the provinces are the totals.

10-6 沿海县生产总值（2010年）
Gross Regional Product of Coastal Counties, 2010

单位：万元 (10 000 yuan)

沿海县 Coastal County		地区生产总值 Gross Regional Product			
			第一产业 Primary Industry	第二产业 Secondary Industry	第三产业 Tertiary Industry
合 计	**Total**	**602 655 353**	**44 248 051**	**305 765 730**	**252 641 474**
天 津	**Tianjin**	**50 301 100**	**81 700**	**34 328 100**	**15 891 200**
滨海新区	Binhai New area	50 301 100	81 700	34 328 100	15 891 200
河 北	**Hebei**	**19 031 500**	**2 763 502**	**8 770 459**	**7 497 539**
丰 南	Fengnan	4 535 543	342 711	2 935 909	1 256 923
滦 南	Luannan	2 578 658	567 954	1 155 875	854 829
乐 亭	Leting	2 394 331	570 649	969 214	854 468
唐 海	Tanghai	746 917	142 396	309 892	294 629
海 港	Haigang	3 793 521	20 404	1 443 049	2 330 068
山海关	Shanhaiguan	329 889	43 111	173 638	113 140
北戴河	Beidaihe	285 749	10 116	53 711	221 922
昌 黎	Changli	1 284 555	461 042	479 931	343 582
抚 宁	Funing	1 261 050	371 484	505 550	384 016
黄 骅	Huanghua	1 602 238	181 292	652 258	768 688
海 兴	Haixing	219 049	52 343	91 432	75 274
辽 宁	**Liaoning**	**29 019 971**	**5 021 835**	**15 839 841**	**8 158 295**
长 海	Changhai	534 482	329 195	63 851	141 436
瓦房店	Wafangdian	6 501 921	712 990	3 960 675	1 828 256
普兰店	Pulandian	5 364 873	765 934	3 230 350	1 368 589
庄 河	Zhuanghe	5 006 982	864 918	2 928 269	1 213 795
东 港	Donggang	3 376 008	490 884	1 629 963	1 255 161
凌 海	Linghai	1 797 092	395 048	996 317	405 727
盖 州	Gaizhou	1 600 113	259 577	766 454	574 082
大 洼	Dawa	1 963 022	454 401	1 163 487	345 134
盘 山	Panshan	1 220 519	339 219	594 348	286 952
绥 中	Suizhong	901 621	286 463	239 444	375 714
兴 城	Xingcheng	753 338	123 206	266 683	363 449

10-6 续表1 continued

沿海县 Coastal County		地区生产总值 Gross Regional Product	第一产业 Primary Industry	第二产业 Secondary Industry	第三产业 Tertiary Industry
江　苏	**Jiangsu**	**36 046 400**	**5 644 900**	**18 076 000**	**12 325 500**
海　安	Hai'an	3 555 700	382 500	1 932 600	1 240 600
如　东	Rudong	3 523 600	451 000	1 882 500	1 190 100
启　东	Qidong	4 300 400	544 900	2 299 800	1 455 700
海　门	Haimen	5 001 000	372 900	3 019 500	1 608 600
赣　榆	Ganyu	2 230 700	366 400	1 095 600	768 700
东　海	Donghai	2 001 400	411 000	912 100	678 300
灌　云	Guanyun	1 501 300	409 900	698 800	392 600
灌　南	Guannan	1 400 800	291 900	703 700	405 200
响　水	Xiangshui	1 352 000	294 400	648 700	408 900
滨　海	Binhai	1 981 600	410 900	848 700	722 000
射　阳	Sheyang	2 446 700	553 700	997 000	896 000
东　台	Dongtai	3 815 400	633 700	1 750 700	1 431 000
大　丰	Dafeng	2 935 800	521 700	1 286 300	1 127 800
浙　江	**Zhejiang**	**73 039 561**	**4 858 037**	**41 751 370**	**26 430 155**
象　山	Xiangshan	2 717 348	432 201	1 326 960	958 187
宁　海	Ninghai	2 787 380	299 317	1 581 106	906 957
余　姚	Yuyao	5 678 820	353 847	3 383 695	1 941 278
慈　溪	Cixi	7 574 167	383 422	4 561 786	2 628 959
奉　化	Fenghua	2 250 690	219 830	1 098 599	932 261
洞　头	Dongtou	343 559	33 408	130 349	179 802
平　阳	Pingyang	2 025 917	109 482	1 018 089	898 346
苍　南	Cangnan	2 547 974	202 978	1 230 840	1 114 156
瑞　安	Rui'an	4 572 171	151 704	2 371 785	2 048 682
乐　清	Yueqing	4 958 356	166 175	3 062 073	1 730 108
海　盐	Haiyan	2 382 942	178 723	1 516 608	687 611
海　宁	Haining	4 558 291	212 693	2 782 991	1 562 607
平　湖	Pinghu	3 405 271	161 009	2 174 157	1 070 106

10-6 续表2 continued

沿海县 Coastal County		地区生产总值 Gross Regional Product	第一产业 Primary Industry	第二产业 Secondary Industry	第三产业 Tertiary Industry
绍　兴	Shaoxing	7 761 092	287 871	4 733 967	2 739 254
上　虞	Shangyu	4 362 648	315 089	2 526 018	1 521 541
岱　山	Daishan	1 280 800	171 962	738 947	369 891
嵊　泗	Shengsi	588 982	112 925	201 976	274 081
玉　环	Yuhuan	3 082 175	197 928	1 959 568	924 679
三　门	Sanmen	1 066 243	161 867	492 167	412 209
温　岭	Wenling	5 814 595	420 519	3 104 014	2 290 062
临　海	Linhai	3 280 140	285 087	1 755 675	1 239 378
福　建	**Fujian**	**46 325 308**	**5 201 443**	**25 461 189**	**15 662 676**
连　江	Lianjiang	1 885 285	670 249	654 157	560 879
罗　源	Luoyuan	1 038 872	192 095	672 261	174 516
平　潭	Pingtan	896 855	264 083	182 631	450 141
福　清	Fuqing	4 773 943	644 414	2 420 089	1 709 440
长　乐	Changle	3 031 299	287 757	1 931 588	811 954
仙　游	Xianyou	1 397 987	211 515	598 108	588 364
惠　安	Hui'an	3 993 724	230 425	2 394 575	1 368 724
石　狮	Shishi	3 702 407	147 192	2 091 349	1 463 866
晋　江	Jinjiang	9 088 816	155 136	5 917 303	3 016 377
南　安	Nan'an	4 822 842	183 887	3 074 160	1 564 795
云　霄	Yunxiao	733 099	210 848	255 632	266 619
漳　浦	Zhangpu	1 499 998	431 686	517 146	551 166
诏　安	Zhao'an	972 210	296 284	347 237	328 689
东　山	Dongshan	784 139	201 122	330 723	252 294
龙　海	Longhai	3 655 009	395 945	2 163 234	1 095 830
霞　浦	Xiapu	938 693	246 471	280 924	411 298
福　安	Fu'an	1 771 671	238 779	961 054	571 838
福　鼎	Fuding	1 338 459	193 555	669 018	475 886
山　东	**Shandong**	**75 788 226**	**7 110 096**	**44 046 571**	**24 631 560**
胶　州	Jiaozhou	5 570 587	393 887	3 197 700	1 979 000

10-6 续表3 continued

沿海县 Coastal County		地区生产总值 Gross Regional Product	第一产业 Primary Industry	第二产业 Secondary Industry	第三产业 Tertiary Industry
即墨	Jimo	5 735 446	478 701	3 133 880	2 122 865
胶南	Jiaonan	5 494 562	433 631	3 244 321	1 816 610
垦利	Kenli	2 202 799	136 694	1 451 453	614 652
利津	Lijin	1 437 881	207 230	794 973	435 678
广饶	Guangrao	4 594 840	313 990	3 330 746	950 104
长岛	Changdao	500 160	298 868	50 725	150 567
龙口	Longkou	6 800 674	272 733	4 377 844	2 150 097
莱阳	Laiyang	3 128 310	337 123	1 896 920	894 267
莱州	Laizhou	4 677 857	497 282	2 755 638	1 424 937
蓬莱	Penglai	3 503 948	225 342	2 075 328	1 203 278
招远	Zhaoyuan	4 529 631	270 744	2 753 648	1 505 239
海阳	Haiyang	2 245 986	460 493	1 050 542	734 951
寿光	Shouguang	4 702 818	670 026	2 440 521	1 592 271
昌邑	Changyi	2 302 867	270 205	1 393 896	638 766
文登	Wendeng	6 005 628	423 914	3 413 682	2 168 032
荣成	Rongcheng	6 362 421	602 006	3 488 222	2 272 193
乳山	Rushan	3 130 985	272 961	1 764 126	1 093 898
无棣	Wudi	1 751 415	284 177	986 936	480 302
沾化	Zhanhua	1 109 411	260 089	445 470	403 853
广东	**Guangdong**	**260 638 004**	**9 200 621**	**113 264 742**	**138 172 641**
越秀	Yuexiu	16 524 043		496 357	16 027 686
荔湾	Liwan	6 147 639	48 176	1 758 999	4 340 464
海珠	Haizhu	7 296 775	31 689	1 332 318	5 932 768
天河	Tianhe	18 722 872	25 814	2 758 188	15 938 870
白云	Baiyun	9 390 914	263 378	2 319 751	6 807 785
黄埔	Huangpu	5 672 896	12 837	3 674 234	1 985 825
番禺	Panyu	10 631 540	453 457	4 386 006	5 792 077
南沙	Nansha	4 882 454	135 991	3 937 997	808 466
萝岗	Luogang	13 816 375	60 670	10 964 105	2 791 600
福田	Futian	18 553 516	7 420	1 958 780	16 587 316
罗湖	Luohu	10 276 679	1 611	957 916	9 317 152

10-6 续表4 continued

沿海县 Coastal County		地区生产总值 Gross Regional Product	第一产业 Primary Industry	第二产业 Secondary Industry	第三产业 Tertiary Industry
盐　田	Yantian	2 820 603	598	756 362	2 063 643
南　山	Nanshan	20 028 401	14 383	12 087 003	7 927 015
宝　安	Bao'an	26 164 770	21 755	17 130 843	9 012 172
龙　岗	Longgang	17 971 132	18 903	12 342 784	5 609 445
香　洲	Xiangzhou	7 554 163	29 444	3 425 402	4 099 317
斗　门	Doumen	1 623 373	220 646	823 012	579 715
金　湾	Jinwan	2 908 422	73 462	2 371 661	463 299
龙　湖	Longhu	1 866 043	46 118	1 084 982	734 943
金　平	Jinping	2 571 078	22 126	1 232 868	1 316 084
濠　江	Haojiang	654 149	58 643	487 378	108 128
潮　阳	Chaoyang	1 956 037	152 622	1 265 424	537 991
潮　南	Chaonan	1 789 744	98 355	1 107 531	583 858
澄　海	Chenghai	2 370 693	225 579	1 375 914	769 200
南　澳	Nan'ao	94 548	29 316	35 369	29 863
蓬　江	Pengjiang	3 874 929	79 146	2 157 000	1 638 783
江　海	Jianghai	1 012 522	32 690	651 983	327 849
新　会	Xinhui	3 950 713	263 321	2 611 315	1 076 077
台　山	Taishan	2 309 696	269 471	1 386 408	653 817
恩　平	Enping	933 095	144 817	333 044	455 234
赤　坎	Chikan	1 447 054	15 915	524 636	906 503
霞　山	Xiashan	2 332 997	17 665	1 199 207	1 116 125
坡　头	Potou	1 991 502	107 659	1 678 269	205 574
麻　章	Mazhang	562 781	115 520	336 692	110 569
遂　溪	Suixi	1 391 238	603 544	368 621	419 073
徐　闻	Xuwen	777 526	390 232	101 405	285 889
廉　江	Lianjiang	1 828 183	618 907	642 527	566 749
雷　州	Leizhou	1 282 601	580 839	196 224	505 538
吴　川	Wuchuan	1 080 359	183 330	467 853	429 176
茂　南	Maonan	1 072 421	150 880	354 880	566 661
茂　港	Maogang	875 934	167 949	330 233	377 752
电　白	Dianbai	2 127 289	586 830	697 970	842 489

10-6 续表5 continued

沿海县 Coastal County		地区生产总值 Gross Regional Product	第一产业 Primary Industry	第二产业 Secondary Industry	第三产业 Tertiary Industry
惠　城	Huicheng	3 424 155	133 120	1 201 841	2 089 194
惠　阳	Huiyang	1 841 256	90 724	939 871	810 661
惠　东	Huidong	2 505 337	283 244	1 265 978	956 115
城　区	Cheng District	699 085	105 850	326 056	267 179
海　丰	Haifeng	1 528 952	241 842	661 031	626 079
陆　丰	Lufeng	1 372 137	324 306	573 536	474 295
江城区	Jiangcheng	1 021 916	146 011	561 478	314 427
阳　西	Yangxi	873 931	350 653	314 955	208 323
阳　东	Yangdong	1 298 872	278 768	678 344	341 760
饶　平	Raoping	1 334 925	242 088	585 982	506 855
揭　东	Jiedong	2 316 004	290 417	1 417 117	608 470
惠　来	Huilai	1 281 735	331 890	629 102	320 743
广　西	**Guangxi**	**1 811 617**	**537 737**	**664 810**	**609 070**
合　浦	Hepu	1 359 864	456 551	498 835	404 478
东　兴	Dongxing	451 753	81 186	165 975	204 592
海　南	**Hainan**	**10 653 666**	**3 828 180**	**3 562 648**	**3 262 838**
琼　海	Qionghai	1 129 861	483 254	167 134	479 473
儋　州	Danzhou	3 117 712	751 808	1 495 753	870 151
文　昌	Wenchang	1 187 277	497 846	262 419	427 012
万　宁	Wanning	980 701	307 672	241 491	431 538
东　方	Dongfang	735 827	230 977	323 838	181 012
澄　迈	Chengmai	1 100 773	345 339	483 240	272 194
临　高	Lingao	717 538	484 835	84 965	147 738
昌　江	Changjiang	608 369	147 804	355 651	104 914
乐　东	Ledong	553 763	349 253	48 264	156 246
陵　水	Lingshui	521 845	229 392	99 893	192 560

注：本表各省数据为合计数。

Note: The data for the provinces are the totals.

10-7 沿海地区财政收支
Financial Revenue and Expenditure by Coastal Regions

单位：亿元 (10 000 million yuan)

地 区 Region	一般预算收入 General Budgetary Revenue	一般预算支出 General Budgetary Expenditure
合 计 **Total**	**29 325.71**	**40 455.49**
天 津 Tianjin	1 455.13	1 796.33
河 北 Hebei	1 737.77	3 537.39
辽 宁 Liaoning	2 643.15	3 905.85
上 海 Shanghai	3 429.83	3 914.88
江 苏 Jiangsu	5 148.91	6 221.72
浙 江 Zhejiang	3 150.80	3 842.59
福 建 Fujian	1 501.51	2 198.18
山 东 Shandong	3 455.93	5 002.07
广 东 Guangdong	5 514.84	6 712.40
广 西 Guangxi	947.72	2 545.28
海 南 Hainan	340.12	778.80

10-8 沿海地区教育基本情况
Basic Conditions of Education by Coastal Regions

地　区 Region	高等学校数 （所） Institutions of Higher Education (unit)	本、专科在校学生数 （人） Number of Students Enrolled in Undergraduate or Specialized Courses (person)	本、专科毕（结）业生数 （人） Number of Students Graduated in Undergraduate or Specialized Courses (person)
全国总计 National Total	**2 483**	**23 085 078**	**6 081 565**
天　津 Tianjin	56	449 702	108 723
河　北 Hebei	119	1 149 252	311 141
辽　宁 Liaoning	114	902 231	236 341
上　海 Shanghai	66	511 283	139 027
江　苏 Jiangsu	156	1 659 415	477 137
浙　江 Zhejiang	103	907 482	238 448
福　建 Fujian	87	674 779	173 702
山　东 Shandong	146	1 645 589	472 882
广　东 Guangdong	134	1 527 254	357 521
广　西 Guangxi	74	600 094	151 052
海　南 Hainan	17	156 700	39 150

10-9 沿海地区卫生基本情况
Basic Conditions of Public Health by Coastal Regions

地　区 Region	卫生机构数 （个） Health Institutions (unit)	医疗机构床位数 （张） Total Beds (bed)	卫生机构人员 （人） Number of Employed Personnel in Health Institutions (person)
全国总计 National Total	**954 389**	**5 159 889**	**8 616 040**
天　津 Tianjin	4 428	49 423	100 169
河　北 Hebei	80 185	266 479	449 108
辽　宁 Liaoning	35 229	215 815	319 116
上　海 Shanghai	4 740	107 130	176 632
江　苏 Jiangsu	31 680	296 390	481 818
浙　江 Zhejiang	30 515	194 759	374 157
福　建 Fujian	27 147	124 232	217 586
山　东 Shandong	68 275	416 148	689 628
广　东 Guangdong	45 930	325 038	626 571
广　西 Guangxi	34 026	152 039	283 543
海　南 Hainan	4 816	28 465	56 893

10-10 沿海地区能源消耗指标
Indicators of Energy Consumption by Coastal Regions

地　区 Region	万元地区生产总值能耗（等价值） Energy Consumption Per 10 000 yuan of GRP (Equivalent Value)		单位工业增加值能耗（规模以上，当量值） Energy Consumption Per 10 000 yuan of Industrial Value-Added (above Designated Size, Equivalent Weight)	万元地区生产总值电耗上升或下降 Electricity Consumption per 10 000 yuan of GRP Change (±%)
	指标值（吨标准煤/万元） Index (ton of SCE/10 000 yuan)	上升或下降（±%） Change (±%)		
天　津 Tianjin	0. 708	-4. 28	-7. 48	-7. 48
河　北 Hebei	1. 300	-3. 69	-6. 68	-0. 36
辽　宁 Liaoning	1. 096	-3. 40	-5. 02	-3. 15
上　海 Shanghai	0. 618	-5. 32	-7. 33	-4. 42
江　苏 Jiangsu	0. 600	-3. 52	-5. 41	-0. 14
浙　江 Zhejiang	0. 590	-3. 07	-2. 40	1. 41
福　建 Fujian	0. 644	-3. 29	-1. 16	2. 73
山　东 Shandong	0. 855	-3. 77	-7. 67	-0. 58
广　东 Guangdong	0. 563	-3. 78	-5. 13	-1. 46
广　西 Guangxi	0. 800	-3. 36	-6. 13	-0. 28
海　南 Hainan	0. 692	5. 23	12. 53	3. 94

注：地区生产总值和工业增加值按2010年价格计算。

Note: Gross regional product and industrial value-added are caculated at 2010 constant prices.

10-11 沿海地区用水情况
Water Use of Coastal Regions

地 区 Region	全年供水总量 （万立方米） Total Annual Volume of Water Supply (10 000 m^3)	人均日生活用水量 （升） Per Capita Daily Consumption of Tap Water for Residential Use (liter)
全国总计 National Total	**5 134 222**	**170.9**
天 津 Tianjin	74 483	128.8
河 北 Hebei	173 061	124.5
辽 宁 Liaoning	266 033	126.2
上 海 Shanghai	311 282	183.6
江 苏 Jiangsu	477 044	212.3
浙 江 Zhejiang	273 591	196.3
福 建 Fujian	137 724	188.2
山 东 Shandong	313 500	129.8
广 东 Guangdong	821 640	241.4
广 西 Guangxi	154 483	241.9
海 南 Hainan	36 758	249.2

10-12 沿海地区全社会固定资产投资
Total Investment in Fixed Assets by the Whole Society of Coastal Regions

单位：亿元 (100 million yuan)

地　区 Region	全社会固定资产投资 Total Investment in Fixed Assets in the Whole Country	#农、林、牧、渔业 Agriculture, Forestry, Animal Husbandry and Fishery	#交通运输、仓储和邮政业 Transport, Storage and Post
全国总计 National Total	**311 485.1**	**8 757.8**	**28 291.7**
天　津 Tianjin	7 067.7	151.2	506.5
河　北 Hebei	16 389.3	590.4	1 433.1
辽　宁 Liaoning	17 726.3	536.9	909.4
上　海 Shanghai	4 962.1	18.6	519.8
江　苏 Jiangsu	26 692.6	225.4	1 225.6
浙　江 Zhejiang	14 185.3	131.6	1 119.2
福　建 Fujian	9 910.9	175.7	1 207.6
山　东 Shandong	26 749.7	731.7	1 456.8
广　东 Guangdong	17 069.2	302.1	1 677.2
广　西 Guangxi	7 990.7	314.3	791.8
海　南 Hainan	1 657.2	24.2	100.3

10-13 沿海地区按经营单位所在地分货物进出口总额
Total Value of Imports and Exports by Location of Importers/Exporters of Coastal Regions

单位：万美元 (10 000 US$)

地区 Region	进出口 Total	出口 Exports	进口 Imports
全国总计 National Total	**364 186 445**	**189 838 089**	**174 348 356**
天津 Tianjin	10 337 617	4 448 194	5 889 423
河北 Hebei	5 360 084	2 856 985	2 503 099
辽宁 Liaoning	9 603 585	5 104 236	4 499 350
上海 Shanghai	43 754 862	20 967 384	22 787 477
江苏 Jiangsu	53 958 089	31 259 006	22 699 084
浙江 Zhejiang	30 937 777	21 634 949	9 302 827
福建 Fujian	14 352 243	9 283 778	5 068 465
山东 Shandong	23 588 608	12 571 257	11 017 351
广东 Guangdong	91 346 733	53 192 657	38 154 076
广西 Guangxi	2 335 597	1 245 776	1 089 821
海南 Hainan	1 275 604	254 162	1 021 442

10-14 沿海地区城镇居民平均每人全年家庭收入和消费性支出
Per Capita Annual Income and Consumption Expenditure of Urban Households by Coastal Regions

单位：元 (yuan)

地　区 Region	总收入 Total Income	可支配收入 Disposable Income	消费性支出 Consumption Expenditure
全国平均水平 National Average	**23 979.20**	**21 809.78**	**15 160.89**
天　津 Tianjin	29 916.04	26 920.86	18 424.09
河　北 Hebei	19 591.91	18 292.23	11 609.29
辽　宁 Liaoning	22 879.77	20 466.84	14 789.61
上　海 Shanghai	40 532.29	36 230.48	25 102.14
江　苏 Jiangsu	28 971.98	26 340.73	16 781.74
浙　江 Zhejiang	34 264.38	30 970.68	20 437.45
福　建 Fujian	27 378.11	24 907.40	16 661.05
山　东 Shandong	24 889.80	22 791.84	14 560.67
广　东 Guangdong	30 218.76	26 897.48	20 251.82
广　西 Guangxi	20 846.11	18 854.06	12 848.37
海　南 Hainan	20 094.18	18 368.95	12 642.75

10-15 沿海地区农村居民家庭人均纯收入和生活消费支出
Per Capita Annual Net Income and Consumption Expenditure of Rural Households by Coastal Regions

单位：元 (yuan)

地 区 Region	纯收入 Net Income	生活消费支出 Consumption Expenditure
全国平均水平 National Average	**6 977.29**	**5 221.13**
天 津 Tianjin	12 321.22	6 725.42
河 北 Hebei	7 119.69	4 711.16
辽 宁 Liaoning	8 296.54	5 406.41
上 海 Shanghai	16 053.79	11 049.32
江 苏 Jiangsu	10 804.95	8 094.57
浙 江 Zhejiang	13 070.69	9 965.08
福 建 Fujian	8 778.55	6 540.85
山 东 Shandong	8 342.13	5 900.57
广 东 Guangdong	9 371.73	6 725.55
广 西 Guangxi	5 231.33	4 210.89
海 南 Hainan	6 446.01	4 166.13

10-16 沿海地区年末人口数
Population at Year-end by Coastal Regions

单位：万人 (10 000 persons)

地 区 Region	2006	2007	2008	2009	2010	2011
全国总计 National Total	**131 448**	**132 129**	**132 802**	**133 450**	**134 091**	**134 735**
天 津 Tianjin	1 705	1 115	1 176	1 228	1 299	1 355
河 北 Hebei	6 898	6 943	6 989	7 034	7 194	7 241
辽 宁 Liaoning	4 271	4 298	4 315	4 341	4 375	4 383
上 海 Shanghai	1 964	2 064	2 141	2 210	2 303	2 347
江 苏 Jiangsu	7 656	7 723	7 762	7 810	7 869	7 899
浙 江 Zhejiang	5 072	5 155	5 212	5 276	5 447	5 463
福 建 Fujian	3 585	3 612	3 639	3 666	3 693	3 720
山 东 Shandong	9 309	9 367	9 417	9 470	9 588	9 637
广 东 Guangdong	9 442	9 660	9 893	10 130	10 441	10 505
广 西 Guangxi	4 719	4 768	4 816	4 856	4 610	4 645
海 南 Hainan	836	845	854	864	869	877

注:2010年数据为当年人口普查数据推算数；其余年份数据为年度人口抽样调查推算数据。各地区数据为常住人口口径。

Note: Data of 2010 are the census year estimates; the rest are the estimates from the annual national sample survey on population changes. Data by region are usual residents.

10-17 沿海城市人口情况（2010年）
Population of Coastal Cities, 2010

单位：万人 (10 000 persons)

沿海城市 Coastal City		年末总人口 Total Population by the End of the Year
合　计	**Total**	**24 257.3**
天　津	**Tianjin**	**984.9**
河　北	**Hebei**	**1 754.2**
唐　山	Tangshan	735.0
秦皇岛	Qinhuangdao	288.3
沧　州	Cangzhou	730.9
辽　宁	**Liaoning**	**1 784.7**
大　连	Dalian	586.4
丹　东	Dandong	241.4
锦　州	Jinzhou	308.3
营　口	Yingkou	235.5
盘　锦	Panjin	131.3
葫芦岛	Huludao	281.8
上　海	**Shanghai**	**1 412.3**
江　苏	**Jiangsu**	**2 076.7**
南　通	Nantong	762.9
连云港	Lianyungang	497.7
盐　城	Yancheng	816.1

10-17 续表1 continued

沿海城市 Coastal City		年末总人口 Total Population by the End of the Year
浙　江	**Zhejiang**	**3 510.4**
杭　州	Hangzhou	689.1
宁　波	Ningbo	574.1
温　州	Wenzhou	786.8
嘉　兴	Jiaxing	341.6
绍　兴	Shaoxing	438.9
舟　山	Zhoushan	96.8
台　州	Taizhou	583.1
福　建	**Fujian**	**2 647.3**
福　州	Fuzhou	645.9
厦　门	Xiamen	180.2
莆　田	Putian	323.5
泉　州	Quanzhou	685.3
漳　州	Zhangzhou	473.9
宁　德	Ningde	338.5
山　东	**Shandong**	**3 392.8**
青　岛	Qingdao	763.6
东　营	Dongying	184.9
烟　台	Yantai	651.1
潍　坊	Weifang	873.8
威　海	Weihai	253.6
日　照	Rizhao	287.9
滨　州	Binzhou	377.9

10-17 续表2 continued

沿海城市	Coastal City	年末总人口 Total Population by the End of the Year
广　东	**Guangdong**	**5 830.9**
广　州	Guangzhou	806.1
深　圳	Shenzhen	259.9
珠　海	Zhuhai	104.7
汕　头	Shantou	524.1
江　门	Jiangmen	392.3
湛　江	Zhanjiang	777.8
茂　名	Maoming	747.2
惠　州	Huizhou	337.3
汕　尾	Shanwei	345.0
阳　江	Yangjiang	282.8
东　莞	Dongguan	181.8
中　山	Zhongshan	149.2
潮　州	Chaozhou	260.9
揭　阳	Jieyang	661.8
广　西	**Guangxi**	**645.7**
北　海	Beihai	166.8
防城港	Fangchenggang	91.2
钦　州	Qinzhou	387.7
海　南	**Hainan**	**217.4**
海　口	Haikou	160.4
三　亚	Sanya	57.0

注：本表各省数据为合计数。

Note: The data for the provinces are the totals.

10-18 沿海县人口情况（2010年）
Population of Coastal Counties, 2010

单位：万人 (10 000 persons)

沿海县	Coastal County	年末总人口 Total Population by the End of the Year
合　计	**Total**	**12 309.47**
天　津	**Tianjin**	**111**
滨海新区	Binhai New area	111
河　北	**Hebei**	**444.77**
丰　南	Fengnan	54.43
滦　南	Luannan	58.07
乐　亭	Leting	49.40
唐　海	Tanghai	14.30
海　港	Haigang	65.03
山海关	Shanhaiguan	14.41
北戴河	Beidaihe	6.94
昌　黎	Changli	55.80
抚　宁	Funing	49.33
黄　骅	Huanghua	53.93
海　兴	Haixing	23.13
辽　宁	**Liaoning**	**666.30**
长　海	Changhai	7.30
瓦房店	Wafangdian	100.00
普兰店	Pulandian	93.00
庄　河	Zhuanghe	90.50
东　港	Donggang	60.80
凌　海	Linghai	53.00
盖　州	Gaizhou	72.90
大　洼	Dawa	40.30
盘　山	Panshan	29.80
绥　中	Suizhong	63.60
兴　城	Xingcheng	55.10
上　海	**Shanghai**	**537.00**
宝　山	Baoshan	88.30
浦东新区	Pudong New District	275.80

10-18 续表1 continued

沿海县 Coastal County		年末总人口 Total Population by the End of the Year
金　山	Jinshan	51.70
奉　贤	Fengxian	52.20
崇　明	Chongming	69.00
江　苏	**Jiangsu**	**1 276.63**
海　安	Hai'an	93.42
如　东	Rudong	104.84
启　东	Qidong	112.05
海　门	Haimen	99.86
赣　榆	Ganyu	112.62
东　海	Donghai	115.10
灌　云	Guanyun	100.26
灌　南	Guannan	76.16
响　水	Xiangshui	61.33
滨　海	Binhai	117.60
射　阳	Sheyang	97.49
东　台	Dongtai	113.36
大　丰	Dafeng	72.54
浙　江	**Zhejiang**	**1 472.65**
象　山	Xiangshan	54.03
宁　海	Ninghai	61.09
余　姚	Yuyao	83.38
慈　溪	Cixi	103.88
奉　化	Fenghua	48.35
洞　头	Dongtou	12.81
平　阳	Pingyang	86.73
苍　南	Cangnan	129.78
瑞　安	Rui'an	119.05
乐　清	Yueqing	124.05
海　盐	Haiyan	37.31
海　宁	Haining	66.03
平　湖	Pinghu	48.70

10-18 续表2 continued

沿海县 Coastal County		年末总人口 Total Population by the End of the Year
绍 兴	Shaoxing	72.20
上 虞	Shangyu	77.64
岱 山	Daishan	19.11
嵊 泗	Shengsi	7.94
玉 环	Yuhuan	41.96
三 门	Sanmen	42.92
温 岭	Wenling	119.29
临 海	Linhai	116.40
福 建	**Fujian**	**1 283.63**
连 江	Lianjiang	63.22
罗 源	Luoyuan	25.54
平 潭	Pingtan	40.11
福 清	Fuqing	127.50
长 乐	Changle	68.51
仙 游	Xianyou	108.09
惠 安	Hui'an	96.14
石 狮	Shishi	31.66
晋 江	Jinjiang	106.58
南 安	Nan'an	150.41
云 霄	Yunxiao	43.65
漳 浦	Zhangpu	84.17
诏 安	Zhao'an	60.20
东 山	Dongshan	20.92
龙 海	Longhai	80.42
霞 浦	Xiapu	53.15
福 安	Fu'an	65.23
福 鼎	Fuding	58.13
山 东	**Shandong**	**1 222.60**
胶 州	Jiaozhou	80.30

10-18 续表3 continued

沿海县	Coastal County	年末总人口 Total Population by the End of the Year
即　墨	Jimo	112.90
胶　南	Jiaonan	83.90
垦　利	Kenli	22.00
利　津	Lijin	29.90
广　饶	Guangrao	49.60
长　岛	Changdao	4.30
龙　口	Longkou	63.40
莱　阳	Laiyang	87.40
莱　州	Laizhou	85.90
蓬　莱	Penglai	45.00
招　远	Zhaoyuan	57.10
海　阳	Haiyang	66.40
寿　光	Shouguang	103.90
昌　邑	Changyi	58.10
文　登	Wendeng	64.50
荣　成	Rongcheng	67.00
乳　山	Rushan	57.30
无　棣	Wudi	44.80
沾　化	Zhanhua	38.90
广　东	**Guangdong**	**4 634.07**
越　秀	Yuexiu	116.97
荔　湾	Liwan	70.93
海　珠	Haizhu	95.28
天　河	Tianhe	77.06
白　云	Baiyun	83.19
黄　埔	Huangpu	19.97
番　禺	Panyu	100.39
南　沙	Nansha	15.41
萝　岗	Luogang	18.90
福　田	Futian	131.96
罗　湖	Luohu	92.45

10-18 续表4 continued

沿海县 Coastal County		年末总人口 Total Population by the End of the Year
盐　田	Yantian	20.91
南　山	Nanshan	108.94
宝　安	Bao'an	450.51
龙　岗	Longgang	232.43
香　洲	Xiangzhou	89.35
斗　门	Doumen	41.62
金　湾	Jinwan	25.19
龙　湖	Longhu	39.64
金　平	Jinping	75.39
濠　江	Haojiang	28.29
潮　阳	Chaoyang	165.93
潮　南	Chaonan	132.87
澄　海	Chenghai	74.61
南　澳	Nan'ao	7.37
蓬　江	Pengjiang	46.95
江　海	Jianghai	15.91
新　会	Xinhui	75.35
台　山	Taishan	98.56
恩　平	Enping	50.12
赤　坎	Chikan	22.82
霞　山	Xiashan	36.42
坡　头	Potou	39.39
麻　章	Mazhang	27.38
遂　溪	Suixi	105.02
徐　闻	Xuwen	72.56
廉　江	Lianjiang	167.57
雷　州	Leizhou	168.93
吴　川	Wuchuan	110.17
茂　南	Maonan	82.84
茂　港	Maogang	49.93
电　白	Dianbai	143.29

10-18 续表5 continued

沿海县	Coastal County	年末总人口 Total Population by the End of the Year
惠　城	Huicheng	76.25
惠　阳	Huiyang	37.46
惠　东	Huidong	83.91
城　区	Cheng District	38.00
海　丰	Haifeng	82.80
陆　丰	Lufeng	175.22
江　城	Jiangcheng	48.48
阳　西	Yangxi	51.47
阳　东	Yangdong	48.67
饶　平	Raoping	101.49
揭　东	Jiedong	128.73
惠　来	Huilai	132.82
广　西	**Guangxi**	**117.99**
东　兴	Dongxing	12.87
合　浦	Hepu	105.12
海　南	**Hainan**	**542.83**
琼　海	Qionghai	49.64
儋　州	Danzhou	106.86
文　昌	Wenchang	58.46
万　宁	Wanning	61.29
东　方	Dongfang	44.29
澄　迈	Chengmai	55.87
临　高	Lingao	49.12
昌　江	Changjiang	26.50
乐　东	Ledong	53.78
陵　水	Lingshui	37.02

注：本表各省数据为合计数。

Note: The data for the provinces are the totals.

10-19 沿海地区城镇单位就业人员情况
Employed Persons in Urban Units by Coastal Regions

单位：万人 (10 000 persons)

地　区 Region	2010	2011
全国总计 National Total	**13 051.5**	**14 413.3**
天　津 Tianjin	205.7	268.2
河　北 Hebei	519.6	555.4
辽　宁 Liaoning	518.1	579.6
上　海 Shanghai	392.9	497.3
江　苏 Jiangsu	763.8	811.3
浙　江 Zhejiang	883.6	995.7
福　建 Fujian	507.1	596.3
山　东 Shandong	956.2	1 050.4
广　东 Guangdong	1 118.5	1 238.2
广　西 Guangxi	316.7	341.6
海　南 Hainan	81.3	85.1

10-20 沿海城市就业人员情况（2010年）
Number of Employed Persons by Coastal Cities, 2010

单位：万人　　(10 000 persons)

沿海城市 Coastal City		就业人员 Number of Employed Persons	城镇就业人员 Urban Employed Persons
合　计	**Total**	**16 375.50**	**3 802.02**
天　津	**Tianjin**	**728.70**	**205.65**
河　北	**Hebei**	**991.50**	**157.47**
唐　山	Tangshan	434.60	84.02
秦皇岛	Qinhuangdao	160.80	29.75
沧　州	Cangzhou	396.10	43.70
辽　宁	**Liaoning**	**1 068.40**	**249.60**
大　连	Dalian	402.20	96.30
丹　东	Dandong	120.80	24.60
锦　州	Jinzhou	163.70	32.20
营　口	Yingkou	139.30	20.80
盘　锦	Panjin	104.90	50.30
葫芦岛	Huludao	137.50	25.40
上　海	**Shanghai**	**1 090.80**	**392.87**
江　苏	**Jiangsu**	**1 114.10**	**149.02**
南　通	Nantong	463.70	63.11
连云港	Lianyungang	302.10	34.00
盐　城	Yancheng	348.30	51.91
浙　江	**Zhejiang**	**2 755.40**	**754.89**
杭　州	Hangzhou	626.30	232.70
宁　波	Ningbo	476.50	140.18
温　州	Wenzhou	558.90	107.92
嘉　兴	Jiaxing	317.60	80.40
绍　兴	Shaoxing	341.80	107.13
舟　山	Zhoushan	66.70	16.78
台　州	Taizhou	367.60	69.78
福　建	**Fujian**	**1 680.90**	**428.42**
福　州	Fuzhou	391.20	105.48
厦　门	Xiamen	149.90	95.33
莆　田	Putian	176.60	28.78

10-20 续表 continued

沿海城市 Coastal City		就业人员 Number of Employed Persons	城镇就业人员 Urban Employed Persons
泉　州	Quanzhou	521.60	142.17
漳　州	Zhangzhou	275.40	40.11
宁　德	Ningde	166.20	16.55
山　东	**Shandong**	**2 143.90**	**426.90**
青　岛	Qingdao	523.90	124.10
东　营	Dongying	115.60	40.10
烟　台	Yantai	416.80	90.50
潍　坊	Weifang	513.80	74.00
威　海	Weihai	154.30	39.90
日　照	Rizhao	180.20	20.70
滨　州	Binzhou	239.30	37.60
广　东	**Guangdong**	**4 319.60**	**911.10**
广　州	Guangzhou	789.10	246.40
深　圳	Shenzhen	705.20	253.00
珠　海	Zhuhai	105.40	63.20
汕　头	Shantou	237.90	32.30
江　门	Jiangmen	248.30	44.90
湛　江	Zhanjiang	318.10	40.70
茂　名	Maoming	298.00	30.90
惠　州	Huizhou	255.40	80.40
汕　尾	Shanwei	123.50	15.70
阳　江	Yangjiang	170.10	18.20
东　莞	Dongguan	438.50	23.20
中　山	Zhongshan	217.80	29.00
潮　州	Chaozhou	146.30	12.50
揭　阳	Jieyang	266.00	20.70
广　西	**Guangxi**	**340.90**	**34.50**
北　海	Beihai	58.10	11.90
防城港	Fangchenggang	55.30	8.90
钦　州	Qinzhou	227.50	13.70
海　南	**Hainan**	**141.30**	**91.60**
海　口	Haikou	111.20	77.20
三　亚	Sanya	30.10	14.40

注：本表各省数据为合计数。

Note: The data for the provinces are the totals.

10-21 沿海县就业人员情况（2010年）
Number of Employed Persons of Coastal Counties, 2010

单位：万人 (10 000 persons)

沿海县 Coastal County		城镇单位职工人数 Urban Employed Persons	乡村从业人员 Rural Laborer
合 计	**Total**	**1 373.85**	**4 276.28**
河 北	**Hebei**	**29.88**	**180.00**
丰 南	Fengnan	5.13	26.53
滦 南	Luannan	2.90	28.98
乐 亭	Leting	2.38	26.50
唐 海	Tanghai	4.43	6.93
海 港	Haigang	2.06	4.59
山海关	Shanhaiguan	1.86	3.20
北戴河	Beidaihe	1.45	2.40
昌 黎	Changli	2.26	28.18
抚 宁	Funing	3.27	24.40
黄 骅	Huanghua	3.03	17.87
海 兴	Haixing	1.11	10.42
辽 宁	**Liaoning**	**51.27**	**279.94**
长 海	Changhai	0.81	3.11
瓦房店	Wafangdian	4.00	37.78
普兰店	Pulandian	5.55	35.70
庄 河	Zhuanghe	3.78	37.43
东 港	Donggang	2.72	26.55
凌 海	Linghai	2.44	23.64
盖 州	Gaizhou	1.68	32.95
大 洼	Dawa	18.51	18.62
盘 山	Panshan	6.94	16.87
绥 中	Suizhong	2.11	27.79
兴 城	Xingcheng	2.73	19.50

10-21 续表1 continued

沿海县 Coastal County		城镇单位职工人数 Urban Employed Persons	乡村从业人员 Rural Laborer
江　苏	**Jiangsu**	**67.68**	**531.51**
海　安	Hai'an	6.65	40.09
如　东	Rudong	6.74	49.55
启　东	Qidong	6.04	54.89
海　门	Haimen	6.70	50.08
赣　榆	Ganyu	3.93	41.92
东　海	Donghai	4.06	47.71
灌　云	Guanyun	4.12	36.43
灌　南	Guannan	3.46	31.43
响　水	Xiangshui	3.27	21.74
滨　海	Binhai	3.76	41.79
射　阳	Sheyang	5.87	33.95
东　台	Dongtai	6.80	50.19
大　丰	Dafeng	6.28	31.74
浙　江	**Zhejiang**	**210.73**	**808.03**
象　山	Xiangshan	26.44	26.88
宁　海	Ninghai	5.66	32.96
余　姚	Yuyao	7.28	45.81
慈　溪	Cixi	9.19	79.86
奉　化	Fenghua	6.57	26.85
洞　头	Dongtou	0.85	5.33
平　阳	Pingyang	7.34	38.14
苍　南	Cangnan	8.55	65.46
瑞　安	Rui'an	13.34	64.24
乐　清	Yueqing	23.31	67.76
海　盐	Haiyan	7.35	21.39
海　宁	Haining	13.12	31.27
平　湖	Pinghu	13.77	22.24

10-21 续表2 continued

沿海县 Coastal County		城镇单位职工人数 Urban Employed Persons	乡村从业人员 Rural Laborer
绍　兴	Shaoxing	19.33	48.94
上　虞	Shangyu	18.91	37.42
岱　山	Daishan	1.69	8.84
嵊　泗	Shengsi	1.03	2.72
玉　环	Yuhuan	5.15	36.90
三　门	Sanmen	2.77	21.68
温　岭	Wenling	9.25	65.47
临　海	Linhai	9.83	57.87
福　建	**Fujian**	**157.51**	**597.37**
连　江	Lianjiang	4.25	29.56
罗　源	Luoyuan	1.34	9.98
平　潭	Pingtan	1.49	18.61
福　清	Fuqing	22.14	54.91
长　乐	Changle	4.09	26.67
仙　游	Xianyou	4.09	49.53
惠　安	Hui'an	23.52	46.91
石　狮	Shishi	7.55	12.66
晋　江	Jinjiang	53.59	63.91
南　安	Nan'an	8.74	78.38
云　霄	Yunxiao	2.28	17.27
漳　浦	Zhangpu	5.90	44.64
诏　安	Zhao'an	2.70	32.03
东　山	Dongshan	1.80	8.05
龙　海	Longhai	7.33	38.09
霞　浦	Xiapu	1.72	21.62
福　安	Fu'an	2.77	18.43
福　鼎	Fuding	2.21	26.12
山　东	**Shandong**	**125.53**	**557.72**
胶　州	Jiaozhou	13.93	36.68

10-21 续表3 continued

沿海县 Coastal County		城镇单位职工人数 Urban Employed Persons	乡村从业人员 Rural Laborer
即　墨	Jimo	13. 39	56. 56
胶　南	Jiaonan	10. 50	36. 32
垦　利	Kenli	2. 62	8. 53
利　津	Lijin	1. 61	14. 37
广　饶	Guangrao	6. 22	25. 94
长　岛	Changdao	0. 52	1. 49
龙　口	Longkou	7. 39	27. 59
莱　阳	Laiyang	7. 34	42. 10
莱　州	Laizhou	9. 45	37. 02
蓬　莱	Penglai	7. 80	20. 52
招　远	Zhaoyuan	6. 03	22. 81
海　阳	Haiyang	3. 45	35. 53
寿　光	Shouguang	7. 47	40. 69
昌　邑	Changyi	3. 34	27. 01
文　登	Wendeng	6. 43	27. 64
荣　成	Rongcheng	8. 81	23. 41
乳　山	Rushan	3. 92	27. 66
无　棣	Wudi	3. 20	24. 29
沾　化	Zhanhua	2. 11	21. 56
广　东	**Guangdong**	**703. 91**	**1 088. 99**
越　秀	Yuexiu	53. 13	
荔　湾	Liwan	12. 77	
海　珠	Haizhu	20. 50	
天　河	Tianhe	28. 26	
白　云	Baiyun	21. 86	
黄　埔	Huangpu	11. 58	
番　禺	Panyu	21. 22	
南　沙	Nansha	8. 74	
萝　岗	Luogang	34. 14	
福　田	Futian	63. 08	
罗　湖	Luohu	36. 01	

10-21 续表4 continued

沿海县 Coastal County		城镇单位职工人数 Urban Employed Persons	乡村从业人员 Rural Laborer
盐 田	Yantian	6.99	
南 山	Nanshan	47.96	
宝 安	Bao'an	50.26	
龙 岗	Longgang	48.72	
香 洲	Xiangzhou	39.28	
斗 门	Doumen	11.86	
金 湾	Jinwan	12.01	
龙 湖	Longhu	6.81	11.81
金 平	Jinping	10.36	5.98
濠 江	Haojiang	2.81	9.79
潮 阳	Chaoyang	3.97	58.37
潮 南	Chaonan	2.89	53.72
澄 海	Chenghai	2.65	33.97
南 澳	Nan'ao	0.51	2.42
蓬 江	Pengjiang	11.80	10.18
江 海	Jianghai	3.31	5.33
新 会	Xinhui	7.40	33.00
台 山	Taishan	4.79	49.28
恩 平	Enping	3.30	18.26
赤 坎	Chikan	5.36	1.83
霞 山	Xiashan	7.74	1.95
坡 头	Potou	2.00	18.30
麻 章	Mazhang	1.48	12.87
遂 溪	Suixi	3.99	43.51
徐 闻	Xuwen	3.77	31.11
廉 江	Lianjiang	6.00	72.00
雷 州	Leizhou	5.48	42.69
吴 川	Wuchuan	4.21	53.29
茂 南	Maonan	1.49	22.23
茂 港	Maogang	1.40	19.08
电 白	Dianbai	5.40	56.29

10-21 续表5 continued

沿海县 Coastal County		城镇单位职工人数 Urban Employed Persons	乡村从业人员 Rural Laborer
惠　城	Huicheng	18.50	30.93
惠　阳	Huiyang	13.91	30.15
惠　东	Huidong	5.96	36.83
城　区	Cheng District	2.43	16.31
海　丰	Haifeng	3.77	39.33
陆　丰	Lufeng	4.60	60.88
江　城	Jiangcheng	1.97	14.11
阳　西	Yangxi	2.48	25.67
阳　东	Yangdong	2.92	25.84
饶　平	Raoping	2.95	44.14
揭　东	Jiedong	3.21	59.75
惠　来	Huilai	3.92	37.79
广　西	**Guangxi**	**4.29**	**37.91**
东　兴	Dongxing	0.90	5.48
合　浦	Hepu	3.39	32.43
海　南	**Hainan**	**23.05**	**194.81**
琼　海	Qionghai	2.60	19.02
儋　州	Danzhou	4.51	31.74
文　昌	Wenchang	2.64	21.29
万　宁	Wanning	2,40	18.17
东　方	Dongfang	2.17	17.17
澄　迈	Chengmai	2.39	23.05
临　高	Lingao	1.48	19.13
昌　江	Changjiang	1.89	8.50
乐　东	Ledong	1.69	23.50
陵　水	Lingshui	1.28	13.24

注：本表各省数据为合计数，缺少福建金门县数据。

Note: The data for the provinces are the totals, and lacking the data for Jinmen of Fujian Province.

主要统计指标解释

1. 国内(或地区)生产总值 指一个国家（或地区）所有常驻单位在一定时期内生产活动的最终成果。国内生产总值有三种表现形态，即价值形态、收入形态和产品形态。从价值形态看，它是所有常驻单位在一定时期内生产的全部货物和服务价值超过同期中间投入的全部非固定资产货物和服务价值的差额，即所有常驻单位的增加值之和；从收入形态看，它是所有常驻单位在一定时期内创造并分配给常驻单位和非常驻单位的初次收入分配之和；从产品形态看，它是所有常驻单位在一定时期内最终使用的货物和服务价值与货物和服务净出口价值之和。在实际核算中，国内生产总值有三种计算方法，即生产法、收入法和支出法。三种方法分别从不同的方面反映国内生产总值及其构成。

2. 三次产业 是根据社会生产活动历史发展的顺序对产业结构的划分，产品直接取自自然界的部门称为第一产业，对初级产品进行再加工的部门称为第二产业，为生产和消费提供各种服务的部门称为第三产业。它是世界上较为通用的产业结构分类，但各国的划分不尽一致。我国的三次产业划分是：

第一产业：是指农、林、牧、渔业。

第二产业：是指采矿业，制造业，电力、燃气及水的生产和供应业，建筑业。

第三产业：是指除第一、二产业以外的其他行业。第三产业包括：交通运输、仓储和邮政业，信息传输、计算机服务和软件业，批发和零售业，住宿和餐饮业，金融业，房地产业，租赁和商务服务业，科学研究、技术服务和地质勘查业，水利、环境和公共设施管理业，居民服务和其他服务业，教育，卫生、社会保障和社会福利业，文化、体育和娱乐业，公共管理和社会组织，国际组织。

3. 增加值 是指各行各业生产经营和劳务活动的最终成果，采用生产法和收入法两种方法计算。

生产法 是从货物和服务活动在生产过程中形成的总产品入手，剔除生产过程中投入的中间产品价值，得到新增价值的方法。

收入法 又称分配法。按收入法计算国内生产总值是从生产过程创造的收入的角度对常驻单位的生产活动成果进行核算；按照此法计算，增加值由劳动者报酬、固定资产折旧、生产税净额和营业盈余四个部分组成。

4. 年末总人口 是指每年 12 月 31 日 24 时一定地区范围内的有生命的个人的人口总和。

5. 财政收入 指国家财政参与社会产品分配所取得的收入，是实现国家职能的财力保证。财政收入所包括的内容几经变化，目前主要包括：

(1)各项税收：包括增值税、营业税、消费税、土地增值税、城市维护建设税、资源税、城市土地使用税、企业所得税、个人所得税、关税、证券交易印花税、车辆购置税、农牧业税和耕地占用税等。

(2)专项收入：包括排污费收入、城市水资源费收入、矿产资源补偿费收入、教育费附加收入等。

(3)其他收入：包括利息收入、基本建设贷款归还收入、基本建设收入、捐赠收入等。

(4)国有企业亏损补贴：此项为负收入，冲减财政收入。主要包括对工业企业、商业企业、粮食企业的补贴。

6. 财政支出 国家财政将筹集起来的资金进行分配使用，以满足经济建设和各项事业的需要。

7. 基本建设支出 指按国家有关规定，属于基本建设范围内的基本建设有偿使用、拨款、资本金支出以及经国家批准对专项和政策性基建投资贷款，在部门的基建投资额中统筹支付的贴息支出。

8. 普通高等学校 指按照国家规定的设置标准和审批程序批准举办的，通过全国普通高等学校统一招生考试，招收高中毕业生为主要培养对象，实施高等教育的全日制大学、独立设置的学院和高等专科学校、高等职业学校和其他机构。

大学、独立设置的学院主要实施本科层次以上教育，高等专科学校、高等职业学校实施专科层次教育，其他机构是承担国家普通招生计划任务不计校数的机构。包括普通高等学校分校和批准筹建的普通高等学校等。

9. 卫生机构 包括医疗机构、疾病预防控制中心(防疫站)、采供血机构、卫生监督及监测(检验)机构、医学科研和在职培训机构、健康教育所等。

10. 医疗机构 包括医院、社区卫生服务中心(站)、疗养院、卫生院、门诊部、诊所(卫生所、医务室)、妇幼保健院(所、站)、专科疾病防治院(所、站)、急救中心(站)和临床检验中心。医疗机构分为非赢利性医疗机构和赢利性医疗机构。

11. 单位国内生产总值能耗 指一定时期内，一个国家或地区每生产一个单位的国内生产总值所消耗的能源。计算公式为：

$$\text{单位国内生产总值能源}=\frac{\text{能源消费总量}}{\text{国内生产总值}}$$

12. 单位国内生产总值电耗 指一定时期内，一个国家或地区每生产一个单位的国内生产总值所消耗的电力。计算公式为：

$$\text{单位国内生产总值电耗}=\frac{\text{全社会用电量}}{\text{国内生产总值}}$$

13. 单位工业增加值能耗 指一定时期内，一个国家或地区每生产一个单位的工业增加值所消耗的能源。计算公式为：

$$\text{单位工业增加值能耗}=\frac{\text{工业能源消费总量}}{\text{工业增加值}}$$

14. 用水总量 指分配给各类用户的包括输水损失在内的毛用水量之和，不包括海水直接利用量。

15. 全社会固定资产投资 是以货币形式表现的在一定时期内全社会建造和购置固定资产的工作量以及与此有关的费用的总称。该指标是反映固定资产投资规模、结构和发展速度的综合性指标，又是观察工程进度和考核投资效果的重要依据。全社会固定资产投资按登记注册类型可分为国有、集体、个体、联营、股份制、外商、港澳台商、其他等。

16. 进出口总额 指实际进出我国国境的货物总金额。包括对外贸易实际进出口货物，来料加工装配进出口货物，国家间、联合国及国际组织无偿援助物资和赠送品，华侨、港澳台同胞和外籍华人捐赠品，租赁期满归承租人所有的租赁货物，进料加工进出口货物，边境地方贸易及边境地区小额贸易进出口货物(边民互市贸易除外)，中外合资企业、中外合作经营企业、外商独资经营企业进出口货物和公用物品，到、离岸价格在规定限额以上的进出口货样和广告品(无商业价值、无使用价值和免费提供出口的除外)，从保税仓库提取在中国境内销售的进口货物，以及其他进出口货物。该指标可以观察一个国家在对外贸易方面的总规模。我国规定出口货物按离岸价格统计，进口货物按到岸价格统计。

17. 商品经营单位所在地进、出口额　指在所在地海关注册登记的有进出口经营权的企业实际进、出口额。

18. 城镇家庭可支配收入　指家庭成员得到可用于最终消费支出和其它非义务性支出以及储蓄的总和，即居民家庭可以用来自由支配的收入。它是家庭总收入扣除交纳的所得税、个人交纳的社会保障支出以及记账补贴后的收入。计算公式为：

可支配收入=家庭总收入-交纳所得税-个人交纳的社会保障支出-记账补贴

19. 城镇家庭消费性支出　指家庭用于日常生活的支出，包括食品、衣着、家庭设备用品及服务、医疗保健、交通和通信、娱乐教育文化服务、居住、杂项商品和服务等八大类支出。

20. 总收入　指调查期内农村住户和住户成员从各种来源渠道得到的收入总和。按收入的性质划分为工资性收入、家庭经营收入、财产性收入和转移性收入。

21. 纯收入　指农村住户当年从各个来源得到的总收入相应地扣除所发生的费用后的收入总和。计算方法：

纯收入=总收入-税费支出-家庭经营费用支出-生产性固定资产折旧-赠送农村亲友支出

纯收入主要用于再生产投入和当年生活消费支出，也可用于储蓄和各种非义务性支出。“农民人均纯收入”按人口平均的纯收入水平，反映的是一个地区或一个农户农村居民的平均收入水平。

22. 人口数　指一定时点、一定地区范围内有生命的个人总和。

年度统计的年末人口数指每年 12 月 31 日 24 时的人口数。年度统计的全国人口总数内未包括香港、澳门特别行政区和台湾省以及海外华侨人数。

23. 城镇人口　城镇人口是指居住在城镇范围内的全部常住人口

24. 就业人员　指在 16 周岁及以上，从事一定社会劳动并取得劳动报酬或经营收入的人员。这一指标反映了一定时期内全部劳动力资源的实际利用情况，是研究我国基本国情国力的重要指标。

Explanatory Notes on Main Statistical Indicators

1. Gross Domestic Product (GDP) refers to the final result of the primary distribution of the income created by all the resident units of a country (or a region) during a certain period of time. Gross domestic product is expressed in three different forms, i.e. value, income, and products respectively. The form of value refers to the total value of all products and services produced by all resident units during a certain period of time minus total value of intermediate input of materials and services of the nature of non-fixed assets or the summation of the value added of all resident units: the form of income includes all the income created by all resident units and distributed primarily to all resident and non-resident units; the form of products refers to the summation of the value of the products and services finally used and the net export value of products and services by all resident units during a given period of time. In the practice of national accounting, gross domestic product is calculated with three approaches, i.e. production approach, income approach, and expenditure approach, which reflect the gross domestic product and its composition from different aspects.

2. Three Industries Industrial structure is classified according to the sequence of historical development of social productive activities. Primary industry refers to the extraction of natural resources; secondary industry involves processing of primary products; and tertiary industry provides services of various kinds for production and consumption. The above classification is universal in the world although it varies to some extent from country to country. The three industries in China are divided as follows:

Primary industry: refers to farming, forestry, sideline production and fishery.

Secondary industry: refers to such industries as mining, manufacturing, production and supply of electric power, fuel gas and water, and construction.

Tertiary industry: refers to all the other industries not included in the primary or secondary industries. It includes such industries as communications and transportation, storage and postal service; information transmission, computer service and software; wholesale and retailing; accommodation and catering; financial service; real estate; charter business and commercial affairs service; scientific research, technological service and geological survey; water conservancy, environmental and other public facilities management; residents service and other service trades; education, health, social security and social welfare; culture, sports and entertainment business as well as public administration and social organizations, and international organizations.

3. Added Value refers to the final result of production operation and labor activities of all trades and professions, which is calculated by using the methods of production and income.

Production Method: refers to the method whereby to get the newly added value by proceeding from the gross product of goods and service activities occurring in the course of production and then rejecting the value of intermediate product input in the course of production.

Income Method: is also called the distribution method. The calculation of the gross domestic product (GDP) by the income method is the accounting of the result of productive activities of permanent units from the angle of the income created in the course of production. According to this method, the added value is composed of the payment for laborers, depreciation for fixed assets, net tax on production and business surplus.

4. Total Population by the End of the Year refers to the sum of living individuals within a particular range of area at 24:00 on December 31 of each year.

5. Government Revenue refers to the income obtained by the government finance through participating in the distribution of social products. It is the financial guarantee to ensure government functioning. The contents of government revenue have changed several times. Now it includes the following main items:

(1) Various tax revenues, including value added tax, business tax, consumption tax, land

value-added tax, tax on city maintenance and construction, resources tax, tax on use of urban land, enterprise income tax, personal income tax, tariff, stamp tax on security transactions, tax on purchase of motor vehicles, tax on agriculture and animal husbandry and tax on occupancy of cultivated land, etc.

(2) Special revenues, including revenues from the fee on sewage treatment, fee on urban water resources, fee for the compensation of mineral resources and extra-charges for education, etc.

(3) Other revenues, including revenues from interest, repayment of capital construction loan, capital construction projects, and donations and grants.

(4) Subsidies for the losses of State-owned enterprises. This is an item of negative revenue, counteracting revenues and consisting of subsidies to industrial, commercial and grain purchasing and supply enterprises.

6. Government Expenditure refers to the distribution and use of the funds which the government finance has raised, so as to meet the needs of economic construction and various causes.

7. Expenditure for capital construction It refers to the non-gratuitous use of, appropriation of funds for and capital outlay on capital construction in the area of capital construction. It also covers the loans on capital construction approved by the government for special purposes or policy purposes and the expenditure with discount paid in an overall way within the amount of the funds appropriated to the departments for capital construction.

8. Regular Institutions of Higher Learning refer to educational establishments set up according to the government evaluation and approval procedures, enrolling graduates from senior secondary schools and providing higher education courses and training for senior professionals. They include full-time universities, colleges, institutions of higher professional education, institutions of higher vocational education and others.

Universities and colleges primarily provide undergraduate courses; institutions of higher professional education and institutions of higher vocational education primarily provide professional trainings; and others refer to educational establishments, which are responsible for enrolling higher education students under the State Plan but not enumerated in the total number of schools, including: branch schools of universities and colleges, and universities and colleges that have been approved and under plan for construction.

9. Health Care Institutions include: medical institutions, disease prevention and control centres (epidemic prevention stations), blood gathering and supplying institutions, health supervision and inspection (check up) institutions, medicinal scientific research and on-job training institutions, health education centres and so on.

10. Medical Organizations include: hospitals, health service centres (stations) in communities, sanatoria, health centres, out-patient clinics, clinics (health stations and infirmaries), maternity and child care agencies (centres and stations), special disease prevention and curing agencies (centres and stations), first aid centres (stations) and clinical inspection centres. Medical organizations are grouped by two types: profit-making and non-profit-making medical organizations.

11. Energy Consumption per Unit of GDP refers to the energy consumption per unit of Gross Domestic Product in a country or the Gross Regional Product in a region in the same reference period. The formula is:

$$\frac{\text{Energy Consumption}}{\text{per Unit of GDP}} = \frac{\text{Total Energy Consumption}}{\text{Gross Domestic Product}}$$

12. Electricity Consumption per Unit of GDP refers to the electricity consumption per unit of Gross Domestic Product in a country or the Gross Regional Product in a region in the same reference period. The formula is:

$$\frac{\text{Electricity Consumption}}{\text{per Unit of GDP}} = \frac{\text{Total Electricity Consumption}}{\text{Gross Domestic Product}}$$

13. Energy Consumption per Unit of Industrial Value-added refers to the energy consumption per unit of industrial value-added in a country or region in the same reference period. The formula is:

$$\frac{\text{Energy Consumption per}}{\text{Unit of Industrial Value - added}} = \frac{\text{Industry Energy Consumption}}{\text{Industrial Value - added.}}$$

14. Gross Amount of Water Used refers to gross water use distributed to users, including loss during transportation, broken down into use by agriculture, industry, living consumption and ecological protection.

15. Total Investment in Fixed Assets in the Whole Country refers to the volume of activities in construction and purchases of fixed assets of the whole country and related fees, expressed in monetary terms during the reference period. It is a comprehensive indicator which shows the size, structure and growth of the investment in fixed assets, providing a basis for observing the progress of construction projects and evaluating results of investment. Total investment in fixed assets in the whole country includes, by type of ownership, the investment by State-owned units, collective-owned units, individuals, joint ownership units, share-holding units, as well as investments by entrepreneurs from foreign countries and from Hong Kong, Macao and Taiwan, and by other units.

16. Total Imports and Exports at Customs refer to the real value of commodities imported and exported across the border of China. They include the actual imports and exports through foreign trade, imported and exported goods under the processing and assembling trades and materials, supplies and gifts as aid given gratis between governments and by the United Nations and other international organizations, and contributions donated by overseas Chinese compatriots in Hong Kong and Macao and Chinese with foreign citizenship, leasing commodities owned by tenant at the expiration of leasing period, the imported and exported commodities processed with imported materials, commodities trading in border areas (excluding mutual exchange goods), the imported and exported commodities and articles for public use of the Sino-foreign joint ventures, cooperative enterprises and ventures with sole foreign investment. Also included is the import or export of samples and advertising goods for which the CIF or FOB value is beyond the permitted ceiling (excluding goods of no trading or use value and free commodities for export), imported goods sold in China from bonded warehouses and other imported or exported goods. The indicator of the total imports and exports at customs can be used to observe the total size of external trade in a country. In accordance with the stipulation of the Chinese government, imports are calculated at CIF, while exports are calculated at FOB.

17. Import-Export Value by Location of China's Foreign Trade Managing Units refers to actual value of imports and exports carried out by corporations which have been registered by the local customs house and are vested with right to run import export business.

18. Disposable Income of Urban Households refers to the actual income at the disposal of members of the households which can be used for final consumption, other non-compulsory expenditure and savings. This equals to total income minus income tax, personal contribution to social security and subsidy for keeping diaries in being a sample household. The following formula is used:

Disposable income = total household income − income tax − personal contribution to social security - subsidy for keeping diaries for a sampled household

19. Consumption Expenditure of Urban Households refers to total expenditure of households for consumption in daily life, including expenditure on the eight categories of food; clothing; household appliances and services; health care and medical services; transport and communications; recreation, education and cultural services; housing; and miscellaneous goods and services.

20. Total Income refers to the sum of income earned from various sources by the rural households and their members during the reference period, and is classified as income from wages and salaries, household operations, properties and transfers.

21. Net Income refers to the total income of rural households from all sources minus all corresponding expenses. The formula for calculation is as follows:

Net income = total income − taxes and fees paid − household operation expenses − taxes and fees − depreciation of fixed assets for production − gifts to non-rural relatives

Net income is mainly used as input for reinvestment in production and as consumption expenditure of the year, and also used for savings and non-compulsory expenses of various forms. "Per capita net income of farmers" is the level of net income averaged by population, reflecting the average income level of rural households in a given area.

22. Total Population refers to the total number of people alive at a certain point of time within a given area.

The annual statistics on total population is taken at midnight, the 3lst of December, not including residents in Taiwan province, Hong Kong and Macao and overseas Chinese.

23. Urban Population refers to all people residing in cities and towns.

24. Employed Persons refers to persons aged 16 and over who are engaged in gainful employment and thus receive remuneration payment or earn business income. This indicator reflects the actual utilization of total labour force during a certain period of time and is often used for the research on China's economic situation and national power.

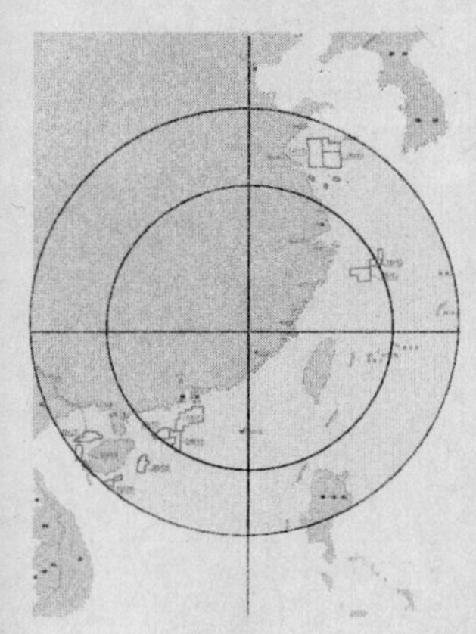

11

部分世界海洋经济统计资料

Part of the World's Marine Economic Statistics Data

11-1　世界海洋面积
The World Ocean Area

区　域 Region	海洋面积（平方公里） Ocean Area (km^2)	占世界海洋面积的比重（%） Proportion in the World Ocean Area (%)	占地球表面面积的比重（%） Proportion in the Earth's Surface Area (%)
合　　计 Total	**361 000 000**	**100.0**	**70.8**
太 平 洋 Pacific Ocean	178 334 000	49.4	35.0
大 西 洋 Atlantic Ocean	91 694 000	25.4	18.0
印 度 洋 Indian Ocean	76 171 000	21.1	14.9
北 冰 洋 Arctic Ocean	14 801 000	4.1	2.9

资料来源：《2011国际统计年鉴》。
Source: *International Statistical Yearbook 2011*.

11-2 主要沿海国家（地区）海岸线长度
Length of Coastline of Major Coastal Countries (Regions)

单位：千米 (km)

国家和地区 Country and Region	海岸线长度 Length of Coastline
美 国 United States	22 680
日 本 Japan	30 000
德 国 Germany	1 300
英 国 United Kingdom	11 450
法 国 France	3 000
意大利 Italy	7 000
加拿大 Canada	20 000
澳大利亚 Australia	20 125
俄罗斯联邦 Russian Fed.	34 000
波 兰 Poland	491
印 度 India	6 083
印度尼西亚 Indonesia	35 000
菲律宾 Philippines	18 533
泰 国 Thailand	3 219
马来西亚 Malaysia	4 675
新加坡 Singapore	193
缅 甸 Myanmar	3 060
孟加拉国 Bangladesh	580
土耳其 Turkey	7 200
韩 国 Republic of Korea	2 413
埃 及 Egypt	2 450
墨西哥 Mexico	9 330
巴 西 Brazil	7 400
阿根廷 Argentina	4 989

11-3 世界主要沿海国家（地区）国土面积和人口（2010年）
Land Area and Population of Major Coastal Countries (Regions), 2010

国家和地区 Country and Region	国土面积(万平方公里) Area of Territory (10 000 km^2)	年中人口（万人） Mid-year Population (10 000 persons)	人口密度（人/平方公里） Population Density (people per km^2)
世界总计 World Total	**13 427.0**	**689 459**	**53**
中 国 China	960.0	133 830	143
美 国 United States	983.2	30 935	34
日 本 Japan	37.8	12 745	350
加拿大 Canada	998.5	3 413	4
德 国 Germany	35.7	8 178	235
英 国 United Kingdom	24.4	6 223	257
法 国 France	54.9	6 490	119
意大利 Italy	30.1	6 048	206
墨西哥 Mexico	196.4	11 342	58
印度尼西亚 Indonesia	190.5	23 987	132
马来西亚 Malaysia	33.1	2 840	86
泰 国 Thailand	51.3	6 912	135
新加坡 Singapore	0.1	508	7 253
韩 国 Korea, Rep.	10.0	4 888	503
印 度 India	328.7	122 462	412
巴 西 Brazil	851.5	19 495	23
俄罗斯联邦 Russian Fed.	1 709.8	14 175	9

资料来源：《2011国际统计年鉴》。

Source: *International Statistical Yearbook 2011*.

11-4 主要沿海国家（地区）国内生产总值
Gross Domestic Product of Major Coastal Countries (Regions)

国家和地区 Country and Region	2010年			2011年
	国内生产总值（亿美元） GDP (100 million US$)	人均国民总收入（美元） GNI per Captia (current US$)	国内生产总值增长率(%) Growth Rate of GDP(%)	国内生产总值（亿美元） GDP (100 million US$)
美国 United States	145 824	47 240	3.0	150 940
日本 Japan	54 978	42 130	4.0	58 695
德国 Germany	33 097	43 290	3.6	35 770
英国 United Kingdom	22 461	38 560	1.4	24 176
法国 France	25 600 *	42 390 *	1.4	27 763
意大利 Italy	20 514	35 150	1.3	21 987
加拿大 Canada	15 741		3.2	17 369
俄罗斯联邦 Russian Fed.	14 798	9 910	4.0	18 504
印度 India	17 290	1 340	10.1	16 761
印度尼西亚 Indonesia	7 066	2 500	6.1	8 457
泰国 Thailand	3 188	4 150	7.8	3 456
马来西亚 Malaysia	2 378	7 760	7.2	2 787
新加坡 Singapore	2 227	41 430	14.5	2 598
韩国 Republic of Korea	10 145	19 890	6.2	11 163
墨西哥 Mexico	10 397	8 930	5.4	11 548
巴西 Brazil	20 879	9 390	7.5	24 929

注：*包括法属圭亚那、瓜德罗普、马提尼克和留尼汪。

Note: *Including French Guiana, Guadeloupe, Martinique and Reunion.

11-5 世界主要沿海国家(地区)国内生产总值产业构成(2010年)
Industrial Composition of GDP of Major Coastal Countries (Regions), 2010

国家和地区 Country and Region	国内生产总值产业构成(%) Industry Structure of GDP(%)		
	第一产业 Primary Industy	第二产业 Secondary Industry	第三产业 Tertiary Industy
世 界 World	**2.9**①	**27.0**①	**70.1**①
美 国 United States	1.2①	21.4①	77.4①
日 本 Japan	1.5①	28.0①	70.5①
德 国 Germany	0.8②	26.5②	72.7②
英 国 United Kingdom	0.7②	21.1②	78.2②
法 国 France	1.8②	19.0②	79.2②
意大利 Italy	1.8②	25.1②	73.1②
墨西哥 Mexico	4.1	34.8	61.1
印度尼西亚 Indonesia	15.9	48.8	35.3
马来西亚 Malaysia	9.5②	44.3②	46.2②
泰 国 Thailand	12.4	44.7	42.9
新加坡 Singapore		28.3	71.7
韩 国 Republic of Korea	2.6②	36.4②	61.0②
印 度 India	16.2	28.4	55.4
巴 西 Brazil	6.0	26.0	68.0
俄罗斯联邦 Russian Fed.	4.7②	32.8②	62.5②

注：①2008年数据。②2009年数据。

Note: ①Data refer to 2008. ②Data refer to 2009.

11-6 世界主要沿海国家（地区）就业人数
Employment in the Major Coastal Countries (Regions)

单位：万人 (10 000 persons)

国家和地区 Country and Region	2000	2005	2007	2008	2009	2010
中　　国 China	72 085	75 825	76 990	77 480	77 995	
中国香港 Hong Kong,China	321	334	319	352	349	352
印度尼西亚 Indonesia	8 984	9 396	9 993	10 449	10 741	10 821
以 色 列 Israel	222	249	268	278	284	297
日　　本 Japan	6 446	6 350	6 412	6 385	6 282	6 256
韩　　国 Korea, Rep.	2 116	2 286	2 343	2 358	2 351	2 383
马来西亚 Malaysia	929	1 007	1 102	1 072	1 102	1 129
菲 律 宾 Philippines	2 745	3 231	3 356	3 409	3 506	3 604
新 加 坡 Singapore			180	185	187	
斯里兰卡 Sri Lanka	631	752	704	718	714	724
泰　　国 Thailand	3 300	3 517	3 625	3 702	3 771	3 804
土 耳 其 Turkey	2 158	2 007	2 079	2 119	2 129	2 261
埃　　及 Egypt	1 720	1 937	2 153	2 251	2 278	2 383
南　　非 South Africa	1 211	1 277	1 347	1 371	1 335	1 306
加 拿 大 Canada	1 476	1 617	1 681	1 709	1 681	1 704
墨 西 哥 Mexico	3 950	4 172	4 306	4 352	4 368	4 414
美　　国 United States	13 689	14 173	14 605	14 536	13 988	13 906
阿 根 廷 Argentina	1 176	944	1 016	1 030	1 037	1 055
巴　　西 Brazil	7 034	1 983	2 044	2 112	2 128	2 202
委内瑞拉 Venezuela	884	1 051	1 132	1 259	1 191	1 200
法　　国 France	2 326	2 498	2 558	2 592	2 565	2 574
德　　国 Germany	3 906	3 887	3 979	4 029	4 031	4 051
意 大 利 Italy	2 121	2 256	2 322	2 341	2 303	2 287
荷　　兰 Netherlands	729	697	731	750	747	739
波　　兰 Poland	1 453	1 412	1 524	1 580	1 587	1 596
俄罗斯联邦 Russian Fed.	6 268	6 817	7 055	7 091	6 938	6 979
西 班 牙 Spain	1 537	1 897	2 036	2 026	1 889	1 846
乌 克 兰 Ukraine	2 018	2 068	2 091	2 097	2 019	2 027
英　　国 United Kingdom	2 748	2 877	2 923	2 944	2 896	2 903
澳大利亚 Australia	904	1 000	1 058	1 087	1 095	1 125
新 西 兰 New Zealand	181	210	217	219	216	219

11-7 主要沿海国家（地区）渔业增加值
Fishery Added Value of Major Coastal Countries (Regions)

单位：亿本币 (100 million local currency units)

国家和地区 Country and Region	渔业增加值 Fishery Added Value					
	2000	2005	2006	2007	2008	2009
孟加拉国 Bangladesh	1 370	1 550	1 630	1 780	1 980	2 180
文莱 Brunei Darsm	...	1		1	...	
柬埔寨 Cambodia	15 200	18 900		24 300		
印度 India	2 150	3 110	3 560	3 910	4 270	
印度尼西亚* Indonesia	30	60	74	98	137	178
伊朗* Iran	2	4	5	5		
韩国* Korea, Rep.	2	2		2	2	
马来西亚 Malaysia	51	56	66	68	75	
菲律宾 Philippines	785	1 160	1 300	1 430	1 700	
泰国 Thailand	1 180	1 040	1 090	1 030	1 060	
越南* Viet Nam	15	33	38	46	58	62
尼日利亚 Nigeria	540	1 700		2 160	2 550	
土耳其 Turkey	4	17		22		
南非 South Africa	6	11				
加拿大 Canada	14	14	13			
墨西哥 Mexico	64	73	62	62		
阿根廷 Argentina	6	16	20	18	24	
委内瑞拉 Venezuela	2 200	9 500	9 880			
法国 France	14	15	14	14	13	11
德国 Germany	2	2	2	2	2	
荷兰 Netherlands	3	2	2	2	2	
波兰 Poland	2	2	2	2	1	
俄罗斯联邦 Russian Fed.		655		616	614	772
西班牙 Spain	15	16	16	17	17	
乌克兰 Ukraine	2	2	2	2	2	
英国 United Kingdom	4	4				
澳大利亚 Australia	251	272	223	267	296	
新西兰 New Zealand	4		2			

注：*万亿本币。

Note: *Trillion local currency units.

11-8 主要沿海国家（地区）旅馆和饭店业增加值
Added Value of Hotel Industry in the Major Coastal Countries (Regions)

单位：亿本币 (100 million local currency units)

国家和地区 Country and Region	旅馆和饭店业增加值 Added Value of Hotel Industry					
	2000	2005	2006	2007	2008	2009
中国 China	2 150	4 200	4 790	5 550	6 620	
中国香港 Hong Kong, China	366	365	414	476		
孟加拉国 Bangladesh	146	251	285	329	389	444
文莱 Brunei Darsm	1	1		1	1	
柬埔寨 Cambodia	5 210	11 200		15 200		
印度 India	2 510	5 440	6 610	7 790	8 140	
印度尼西亚* Indonesia	40	93		124	140	158
伊朗* Iran	2	8	9	11		
哈萨克斯坦 Kazakhstan	148	681	842	1 140	1 320	1 360
韩国* Korea, Rep.	15	18	19	21	23	
马来西亚 Malaysia	80	112	124	142	163	
菲律宾 Philippines	630	959	1 050	1 190	1 300	
新加坡 Singapore	34	38	43	50	55	50
泰国 Thailand	2 750	3 470	3 470	4 170	4 380	
越南* Viet Nam	14	29	36	45	65	75
埃及 Egypt	66	177				
尼日利亚 Nigeria	65	461		728	861	

11-8 续表 continued

国家和地区 Country and Region	旅馆和饭店业增加值 Added Value of Hotel Industry					
	2000	2005	2006	2007	2008	2009
南非 South Africa		140				
加拿大 Canada	234	283	296			
墨西哥 Mexico	2 470	2 380	2 500	2 650		
美国 United States	2 610	3 340	3 580	3 800		
阿根廷 Argentina	78	126	163	200	251	
巴西 Brazil	182	300		417		
委内瑞拉 Venezuela	14 000	55 500	80 800			
法国 France	302	367	383	405	408	416
德国 Germany	301	331	337	370	380	
荷兰 Netherlands	76	85	88	93	90	
波兰 Poland	83	107	111	119	138	
俄罗斯联邦 Russian Fed.		1 710		2 860	3 590	3 390
西班牙 Spain	434	610	652	681	722	
乌克兰 Ukraine	8	24	54	68	97	
英国 United Kingdom	260	327				
澳大利亚 Australia	183	238	256	269	285	
新西兰 New Zealand	21		31			

注：*万亿本币。

Note: *Trillion local currency units.

11-9 主要沿海国家（地区）运输、仓储和通讯业增加值

Added Values of Transportation,Storage and Communications Industries in the Major Coastal Countries (Regions)

单位：亿本币 (100 million local currency units)

国家和地区 Country and Region	运输、仓储和通讯业增加值 Added Values of Transportation, Storage and Commnuications Industries					
	2000	2005	2006	2007	2008	2009
中国 China	6 160	10 700	12 200	14 600	16 400	
中国香港 Hong Kong,China	1 190	1 350	1 370	1 420	1 220	
孟加拉国 Bangladesh	1 970	3 830	4 320	4 890	5 690	6 520
文莱 Brunei Darsm	4	5		5	5	
柬埔寨 Cambodia	9 300	19 000		24 200		
印度 India	15 000	28 300	32 500	36 700	42 000	
印度尼西亚* Indonesia	65	181	232	264	312	352
伊朗* Iran	48	148	191	240		
以色列 Israel	342	417	406	450	465	
日本* Japan	35	34	34	34		
韩国* Korea, Rep	37	54	538 143	59	61	
马来西亚 Malaysia	249	360	389	426	458	
巴基斯坦 Pakistan	4 010	7 600	9 080	10 100	11 800	16 100
菲律宾 Philippines	1 990	4 140	4 460	4 780	5 090	
新加坡 Singapore	208	280	289	334	343	319
斯里兰卡 Sri Lanka	1 350	2 910	3 490	4 280	5 370	
泰国 Thailand	3 960	5 200	5 690	6 260	6 430	
越南* Viet Nam	17	37	44	51	67	73

11-9 续表 continued

国家和地区 Country and Region	运输、仓储和通讯业增加值 Added Values of Transportation, Storage and Commnuications Industries					
	2000	2005	2006	2007	2008	2009
土耳其 Turkey	203	891		1 170		
埃及 Egypt	257	574	580	767	924	
尼日利亚 Nigeria	1 310	4 270		7 200	7 320	
南非 South Africa	809	1 390	1 540	1 640	1 880	2 060
加拿大 Canada	702	918	981			
墨西哥 Mexico	5 570	8 230	9 240	10 100		
美国 United States	6 360	7 390	7 720	8 190		
阿根廷 Argentina	241	444	536	641	803	
巴西 Brazil	866	1 650	1 760	1 980		
委内瑞拉 Venezuela	52 900	16 400	237 000			
法国 France	776	998	1 030	1 080	1 120	1 130
德国 Germany	1 020	1 160	1 220	1 230	1 280	
荷兰 Netherlands	266	332	339	351	350	
波兰 Poland	434	627	685	722	783	
俄罗斯联邦 Russian Fed.	5 930	19 300	22 800	27 500	32 100	32 600
西班牙 Spain	418	563	601	641	663	
乌克兰 Ukraine	198	474	561	701	871	1 040
英国 United Kingdom	693	811				
澳大利亚 Australia	590	804	899	949	1 060	
新西兰 New Zealand	82		117			

注：*万亿本币。
Note: *Trillion local currency units.

11-10 主要沿海国家（地区）海洋渔区标称渔获量
Nominal Catches of Marine Fishing Zone in the Major Coastal Countries (Regions)

国家和地区 Country and Region	2009	2010
世界总计 World Total	**79 247 345**	**77 392 626**
阿尔巴尼亚 Albania	3 694	3 103
阿尔及利亚 Algeria	127 513	93 607
安哥拉 Angola	266 415	250 000 *
安提瓜和巴布达 Antigua Barb	2 490	2 293
阿根廷 Argentina	845 717	796 304
澳大利亚 Australia	170 061	170 566
巴哈马 Bahamas	9 018	11 611
巴林 Bahrain	16 358	13 490
孟加拉国 Bangladesh	602 642	607 492
巴巴多斯 Barbados	3 496	3 269
比利时 Belgium	21 211	21 907
伯利兹 Belize	11 114	114 292
贝宁 Benin	8 678	9 441
巴西 Brazil	585 919	537 247
文莱 Brunei Darsm	1 766	2 272
保加利亚 Bulgaria	7 390	9 684
柬埔寨 Cambodia	75 000	85 094
喀麦隆 Cameroon	65 000 *	65 000 *
加拿大 Canada	917 475	900 329
佛得角群岛 Cape Verde	16 828	19 500 *
智利 Chile	3 453 786	2 679 736
中国台湾 China, Taiwan	769 748	851 307
哥伦比亚 Colombia	83 635	57 978
科摩罗群岛 Comoros	20 450 *	52 261 *
刚果 Congo Rep.	32 833	34 686
库克群岛 Cook Is.	2 695	10 019
哥斯达黎加 Costa Rica	20 750 *	20 750 *
科特迪瓦 Côte dIvoire	38 519	64 749
古巴 Cuba	26 317	21 923
塞浦路斯 Cyprus	1 385	1 400
丹麦 Denmark	777 707	827 975
多米尼加共和国 Dominican Rp.	13 801	14 140
厄瓜多尔 Ecuador	485 467	391 469
埃及 Egypt	127 821	121 362
萨尔瓦多 El Salvador	27 527	33 674

11-10 续表1 continued

国家和地区 Country and Region	2009	2010
赤道几内亚 Eq. Guinea	6 669	6 376 *
爱沙尼亚 Estonia	94 523	92 530
福克兰群岛 Falkland Is.	62 746	99 559
法罗群岛 Faeroe Is.	351 018	393 875
斐济 Fiji	37 320 *	39 671 *
芬兰 Finland	125 327	129 845
法国 France	418 276	424 014
波利尼西亚 Fr. Polynesia	13 321	12 962
加蓬 Gabon	22 000 *	22 000 *
冈比亚 Gambia	41 420	41 970
格鲁吉亚 Georgia	25 000 *	30 544 *
德国 Germany	228 543	207 761
加纳 Ghana	233 115	261 205
希腊 Greece	82 388	82 060 *
格陵兰岛 Greenland	197 878	209 446
格林纳达 Grenada	2 615	2 452
瓜德罗普 Guadeloupe	10 000 *	10 000 *
危地马拉 Guatemala	17 634	19 499
几内亚 Guinea	82 127 *	93 849 *
几内亚比绍 Guinea Bissau	6 650 *	6 650 *
圭亚那 Guyana	42 805	44 386
海地 Haiti	8 000 *	8 000 *
洪都拉斯 Honduras	11 202 *	11 007 *
冰岛 Iceland	1 141 634	1 060 380
印度 India	3 142 802	3 226 211
印度尼西亚 Indonesia	4 807 867	5 035 294
伊朗 Iran	348 122	368 505
伊拉克 Iraq	12 246	13 490
爱尔兰 Ireland	268 997	318 826
以色列 Israel	2 311 *	2 186 *
意大利 Italy	248 013	230 249
牙买加 Jamaica	15 894	15 040
日本 Japan	4 074 569 *	4 004 274 *
肯尼亚 Kenya	5 564	8 264
基里巴斯 Kiribati	40 623	44 599
朝鲜 Korea D. P. Rp.	200 000 *	200 000 *
韩国 Korea Rep.	1 846 865	1 722 672

11-10 续表2 continued

国家和地区 Country and Region	2009	2010
拉脱维亚 Latvia	162 887	164 489
黎巴嫩 Lebanon	3 541 *	3 541 *
利比里亚 Liberia	7 250 *	7 250 *
利比亚 Libya	52 110	50 000 *
立陶宛 Lithuania	171 377	148 541
马达加斯加 Madagascar	98 475	93 336
马来西亚 Malaysia	1 393 214	1 428 882
马尔代夫 Maldives	1 170 661	94 953
马耳他 Malta	1 595	1 836
马尔提尼克 Martinique	6 200 *	5 000 *
毛里塔尼亚 Mauritania	201 900	261 238
毛里求斯 Mauritius	7 676	7 786
墨西哥 Mexico	1 499 383	1 407 479
密克罗尼西亚 Micronesia	27 701	30 866 *
摩洛哥 Morocco	1 *	1 *
莫桑比克 Mozambique	114 383	118 893
缅甸 Myanmar	1 867 510	2 060 780
纳米比亚 Namibia	367 548	367 200 *
荷兰 Netherlands	380 242	387 306
荷属安的列斯群岛 Neth Antilles	18 874	18 455
新喀里多尼亚 New Caledonia	3 554	3 771
新西兰 New Zealand	438 649	435 340
尼加拉瓜 Nicaragua	34 795	36 480
尼日利亚 Nigeria	312 439	323 599
挪威 Norway	2 523 775	2 674 587
阿曼 Oman	158 551	163 927
巴基斯坦 Pakistan	334 007	337 916
帛琉群岛 Palau Is.	1 000 *	1 000 *
巴拿马 Panama	220 607	161 475
巴布亚新几内亚 Papua New Guinea	216 119	211 007
秘鲁 Peru	6 868 732	4 216 723
菲律宾 Philippines	2 416 097	2 426 314
波兰 Poland	205 095	171 213
葡萄牙 Portugal	198 813	222 944
波多黎各 Puerto Rico	1 702	1 898
卡塔尔 Qatar	14 064	13 760
留尼汪岛 Reunion	3 050	3 050 *

11-10 续表3 continued

国家和地区 Country and Region	2009	2010
俄罗斯 Russian Fed.	3 579 992	3 806 641
萨摩亚群岛 Samoa Is.	13 278	12 999 *
圣多美和普林西比 Sao Tome & Principe	4 550 *	4 650 *
沙特阿拉伯 Saudi Arabia	67 664	65 142
塞内加尔 Senegal	406 595	375 414
塞舌尔 Seychelles	81 096	87 108
塞拉利昂 Sierra Leone	186 000 *	186 000 *
新加坡 Singapore	2 121	1 732
索罗门群岛 Solomon Is.	27 941 *	35 179 *
索马里 Somalia	29 800 *	29 800 *
南非 South Africa	511 384	623 020
西班牙 Spain	912 124	962 662
斯里兰卡 Sri Lanka	309 134	383 945
苏丹 Sudan	5 690	5 700 *
苏里南 Suriname	25 411	33 842
瑞典 Sweden	201 851	210 667
叙利亚 Syria	3 107	2 956
坦桑尼亚 Tanzania	46 802	49 438
泰国 Thailand	1 663 846	1 617 399
多哥 Togo	22 025	22 535
汤加 Tonga	2 036 *	2 150 *
特立尼达和多巴哥 Trinidad & Tobago	13 856	13 931
突尼斯 Tunisia	96 668	96 620
土耳其 Turkey	425 046	445 680
特克斯和凯科斯群岛 Turks & Caicos	6 803	5 446
乌克兰 Ukraine	207 318	181 381
阿拉伯联合酋长国 United Arab Em.	77 705	79 610
英国 United Kingdom	588 502	610 182
美国 United States	4 199 378	4 346 719
乌拉圭 Uruguay	80 660	73 105
瓦努阿图 Vanuatu	145 278	97 807
委内瑞拉 Venezuela	263 662 *	249 617 *
越南 Viet Nam	2 091 700	2 226 600
也门 Yemen	159 000 *	191 100

注:*为联合国粮农组织估算值 (表11-11同)。

资料来源：《渔业统计年鉴》，联合国粮农组织，2011年 (表11-11同)。

Note: * It is estimated by FAO from available sources of information or calculation. (Same as Table 11-11).

Source: *Yearbook of Fishery*, FAO, 2011 (Same as Table 11-11).

11-11 主要沿海国家（地区）海藻和其他养殖品种捕捞产量
Capture Production of Seaweeds and Other Aquatic Plants in the Major Coastal Countries (Regions)

单位：吨 (t)

国家和地区 Country and Region	2006	2007	2008	2009	2010
世界总计 World Total	**1 039 465**	**1 074 869**	**1 047 289**	**932 613**	**885 650**
澳大利亚 Australia	15 504	2 223	1 923	1 923 *	1 923 *
加拿大 Canada	43 191	19 382	17 715	43 300	37 632
智利 Chile	301 115	313 551	384 563	368 032	368 580
爱沙尼亚 Estonia	394	1 608	1 483	1 032	351
斐济 Fiji	250 *	200 *	150	150	75
法国 France	19 160	39 757	39 757 *	18 907	22 597
冰岛 Iceland	20 964	21 867	22 559	22 563	21 014
印度尼西亚 Indonesia	4 996	4 643	2 917	3 030	2 697
意大利 Italy	1 400 *	1 400 *	1 400 *	1 400 *	1 400 *
日本 Japan	113 665	103 602	104 668	104 103	96 600 *
韩国 Korea Rep.	13 754	18 189	13 866	10 843	13 043
马达加斯加 Madagascar	1 000	800	800	800	800
墨西哥 Mexico	4 532	5 093	4 900	5 152	1 128
摩洛哥 Morocco	14 870	12 373	9 037	10 368	7 405
挪威 Norway	145 429	134 671	...	...	...
秘鲁 Peru	3 434	10 786	13 779	5 677	4 368
菲律宾 Philippines	314	351	388	434	473
葡萄牙 Portugal	765	495	198	351	498
俄罗斯联邦 Russian Fed.	11 614	8 342	10 242	5 828	5 917
南非 South Africa	9 776	12 472	11 767	11 781	13 007
西班牙 Spain	485	109	97	64	124
坦桑尼亚 Tanzania	278	214	277	280 *	287
乌克兰 Ukraine	1 121	1 892	1 947	2 179	142
美国 United States	6 362	2 272	6 951	8 207	9 027

11-12 主要沿海国家（地区）油气产量（2010年）
Natural Gas and Crude Oil Production of the Major Coastal Countries (Regions), 2010

国家和地区 Country and Region	天然气产量（万亿焦耳） Production of Natural Gas (terajoule)	原油产量（万吨） Production of Crude Oil (10 000 tons)
俄罗斯联邦 Russian Fed.	21 951 492	50 472
美国 United States	23 439 324	37 371
中国 China	3 642 768	20 383
委内瑞拉 Venezuela		14 475
墨西哥 Mexico	3 228 876	13 500
荷兰 Netherlands	2 633 940	102
加拿大 Canada	5 565 756	13 388
尼日利亚 Nigeria		12 008
阿拉伯联合酋长国 United Arab Em.		11 165
挪威 Norway	4 274 952	9 132
英国 United Kingdom	2 391 672	5 803
阿尔及利亚 Algeria		5 405
印度尼西亚 Indonesia	3 593 064	4 601
印度 India		3 670
马来西亚 Malaysia	2 448 924	3 026
厄瓜多尔 Ecuador		2 489
澳大利亚 Australia	1 604 112	2 378
越南 Viet Nam	364 932	1 493
丹麦 Denmark	177 840	1 212
泰国 Thailand		1 207
秘鲁 Peru		841
意大利 Italy	316 152	514
德国 Germany		480
突尼斯 Tunisia		374
乌克兰 Ukraine	672 240	349
巴基斯坦 Pakistan	1 490 580	318
土耳其 Turkey		260
新西兰 New Zealand		238
波兰 Poland	171 552	
法国 France		90
日本 Japan	139 296	74
以色列 Israel	124 728	
西班牙 Spain		12
智利 Chile		12
立陶宛 Lithuania		12

11-13 主要沿海国家（地区）能源利用效率
Use Efficiency of Energy Resources in the Major Coastal Countries (Regions)

单位：吨标准油/万美元 (ton of oil equivalent per 10 000 US$)

国家和地区 Country and Region	单位国内生产总值能耗 Energy consumption per unit of GDP					
	2000	2004	2005	2006	2007	2008
世界 World	**3.02**	**3.04**	**3.00**	**2.96**	**2.92**	**2.94**
中国 China	9.13	9.12	8.88	8.61	7.99	7.86
中国香港 Hong Kong,China	0.79	0.66	0.61	0.60	0.61	0.59
孟加拉国 Bangladesh	3.95	3.89	3.89	3.88	3.80	3.78
文莱 Brunei Darsm	4.09	3.92	3.82	4.69	4.78	5.32
柬埔寨 Cambodia	10.89	9.21	8.37	7.84	7.36	7.02
印度 India	9.98	8.80	8.33	8.00	7.69	7.65
印度尼西亚 Indonesia	9.42	8.94	8.63	8.27	8.17	8.03
伊朗 Iran	11.75	12.61	12.37	12.74	12.79	13.01
以色列 Israel	1.46	1.49	1.46	1.42	1.40	1.37
日本 Japan	1.11	1.07	1.05	1.02	0.99	0.96
韩国 Republic of Korea	3.48	3.26	3.16	3.06	3.02	3.02
马来西亚 Malaysia	5.04	4.86	5.25	5.06	5.19	5.21
巴基斯坦 Pakistan	8.61	8.43	8.10	7.95	7.96	7.70
菲律宾 Philippines	5.05	4.10	3.91	3.69	3.51	3.47
新加坡 Singapore	1.95	1.75	1.97	1.79	1.37	1.27
斯里兰卡 Sri Lanka	5.10	4.71	4.54	4.25	4.06	3.70
泰国 Thailand	5.89	6.27	6.18	6.04	5.98	6.02
土耳其 Turkey	2.86	2.63	2.53	2.61	2.68	2.63

11-13 续表 continued

国家和地区 Country and Region	单位国内生产总值能耗 Energy consumption per unit of GDP					
	2000	2004	2005	2006	2007	2008
越南 Viet Nam	11.89	12.09	11.43	10.88	10.71	10.64
埃及 Egypt	4.52	4.92	5.12	5.03	4.95	4.86
尼日利亚 Nigeria	19.77	17.32	17.05	16.00	15.05	14.98
南非 South Africa	8.30	8.44	7.85	7.61	7.51	7.26
加拿大 Canada	3.47	3.35	3.31	3.18	3.15	3.07
墨西哥 Mexico	2.49	2.58	2.67	2.56	2.55	2.57
美国 United States	2.30	2.13	2.08	2.01	2.00	1.96
阿根廷 Argentina	2.14	2.34	2.14	2.15	1.99	1.94
巴西 Brazil	2.93	2.93	2.91	2.90	2.89	2.90
委内瑞拉 Venezuela	4.82	4.68	4.48	4.17	4.03	3.87
法国 France	1.90	1.91	1.88	1.81	1.75	1.77
德国 Germany	1.78	1.77	1.73	1.69	1.60	1.60
意大利 Italy	1.56	1.59	1.60	1.56	1.51	1.50
荷兰 Netherlands	1.90	1.96	1.92	1.81	1.82	1.77
波兰 Poland	5.20	4.75	4.63	4.59	4.29	4.12
俄罗斯联邦 Russian Fed.	23.84	19.69	18.64	17.73	16.38	15.90
西班牙 Spain	2.10	2.11	2.08	1.99	1.96	1.87
乌克兰 Ukraine	42.80	32.67	31.59	28.30	26.23	25.46
英国 United Kingdom	1.51	1.36	1.33	1.27	1.19	1.18
澳大利亚 Australia	2.59	2.40	2.45	2.43	2.39	2.40
新西兰 New Zealand	3.23	2.75	2.61	2.61	2.55	2.62

11-14 主要沿海国家（地区）国际海运装货量和卸货量
International Ocean Shipping Loading and Unloading Capacities in the Major Coastal Countries (Regions)

单位：万吨 (10 000 tons)

国家和地区 Country and Region	国际海运装货量 International Ocean Shipping Loading Capacity			国际海运卸货量 International Ocean Shipping Unloading Capacity		
	2000	2005	2010	2000	2005	2010
中国香港 Hong Kong,China	6 770	8 918	11 346	10 693	14 095	15 428
孟加拉国 Bangladesh	89	79	466	1 408		3 860
文莱 Brunei Darsm	10	8		102	168	
印度尼西亚 Indonesia	14 153	27 372	50 118	4 504	8 479	11 254
伊朗 Iran	3 065	3 446	4 366①	4 486	6 073	8 232①
以色列 Israel	1 387	1 663	1 927	2 920	2 107	2 414
日本 Japan	13 010			80 654		
韩国 Korea Rep.	15 078	24 250		41 882	51 245	
马来西亚 Malaysia	5 483	7 940	11 240	6 922	10 391	13 682
巴基斯坦 Pakistan	617	1 126	1 950	3 080	3 955	4 870
新加坡② Singapore	32 618	42 266	47 146①			
斯里兰卡 Sri Lanka	919	1 315				
埃及 Egypt		2 123			4 241	
南非 South Africa	2 598					
美国 United States	34 334	40 534		83 352	94 160	
阿根廷③ Argentina	1 550					2 374
法国 France	6 810	10 066	9 986	20 273	22 710	19 931
德国 Germany	8 602	10 832	10 230	14 725	16 866	17 070
荷兰 Netherlands	9 940	12 250	15 335①	32 507	36 424	35 617①
波兰 Poland	3 152	3 835	3 014	1 582	1 642	3 102
俄罗斯联邦 Russian Fed.	828	910	5 258①	84	74	865①
西班牙 Spain	5 627	8 195		19 343	26 164	
乌克兰 Ukraine	4 271	7 070	8 371	684	1 333	1 744
澳大利亚 Australia	48 750	62 401	88 736	5 418	6 989	8 896
新西兰 New Zealand	2 214	2 167	3 040	1 379	1 844	1 798

注：①2009年数据。②包括装货量和卸货量。③不包括转口。

Note: ①Data for 2009. ② Including both goods loaded and unloaded.③Excluding re-exports.

11-15 世界主要外贸货物海运量及构成*（2010年）
World Major Maritime Freight Traffic in Foreign Trade, 2010

品 种 Sort	海运量（百万吨） Freight Traffic (million tons)	
	2010	所占比例（%） Percentage
合 计 Total	**8 367**	**100**
原 油 Crude Oil	1 964	23.0
成品油 Refined Oil	787	9.0
燃 气 Fuel gas	236	3.0
铁矿石 Ironstone	979	12.0
煤 炭 Coal	914	11.0
谷 物 Corn	342	4.0
其他货物 Others	3 144	38.0

注：*为估计数。
资料来源：Shipping Statistics and Market Review, January/February 2011, ISL
Note: *Estimated data.
Source: *Shipping Statistics and Market Review*, January/February 2011, ISL.

11-16 主要沿海国家（地区）国际旅游人数

Number of International Tourists of Major Coastal Countries (Regions)

单位：万人 (10 000 persons)

国家和地区 Country and Region	入境（过夜）旅游人数 Number of Arrivals			出境旅游人数 Number of Departures		
	2000	2005	2009	2000	2005	2009
世界总计 World Total	**69 021**	**80 987**	**89 309**	**73 059**	**88 253**	**96 154**
中国 China	3 123	4 681	5 088	1 047	3 103	4 766
中国香港 Hong Kong, China	881	1 477	1 693	461	7 230	8 196
中国澳门 Macao, China	520	901	1 040	14	30	21
孟加拉国 Bangladesh	20	21	27	113	177	225
文莱 Brunei Darsm	98	13	16			
柬埔寨 Cambodia		133	205	4	57	34
印度 India	265	392	511	442	719	1 107
印度尼西亚 Indonesia	506	500	632	221	411	505
伊朗 Iran	134	189		229		
以色列 Israel	242	190	232	353	369	401
日本 Japan	476	673	679	1 782	1 740	1 545
韩国 Korea, Rep.	532	602	782	551	1 008	949
马来西亚 Malaysia	1 022	1 643	2 365	3 053		
缅甸 Myanmar	21	23	24			
巴基斯坦 Pakistan	56	80				
菲律宾 Philippines	199	262	302	167	214	
新加坡 Singapore	606	708	749	444	516	696
斯里兰卡 Sri Lanka	40	55	45	52	73	96

11-16 续表 continued

国家和地区 Country and Region	入境（过夜）旅游人数 Number of Arrivals			出境旅游人数 Number of Departures		
	2000	2005	2009	2000	2005	2009
泰国 Thailand	958	1 157	1 415	191	305	454
越南 Viet Nam	214	348	375			
土耳其 Turkey	959	2 027	2 551	528	825	1 049
埃及 Egypt	512	824	1 191	296	531	
尼日利亚 Nigeria	81	101				
南非 South Africa	587	737	993	383		442
加拿大 Canada	1 963	1 877	1 574	1 918	2 110	
墨西哥 Mexico	2 064	2 192	2 145	1 108	1 331	1 394
美国 United States	5 124	4 921	5 488	6 133	6 350	6 142
阿根廷 Argentina	291	382	433	495	389	498
巴西 Brazil	531	536	480	323	347	495
委内瑞拉 Venezuela	47	71	62	95	107	165
法国 France	7 719	7 499	7 680	1 989	2 480	
德国 Germany	1 898	2 150	2 422	7 440	7 740	7 230
意大利 Italy	4 118	3 651	4 324	2 199	2 480	2 906
荷兰 Netherlands	1 000	1 001	992	1 390	1 704	1 841
波兰 Poland	1 740	1 520	1 189	5 668	4 084	
俄罗斯联邦 Russian Fed.	2 117	2 220		1 837	2 842	
西班牙 Spain	4 640	5 591	5 223	410	1 046	1 284
乌克兰 Ukraine	643	1 763	2 080	1 342	1 645	1 533
英国 United Kingdom	2 321	2 804	2 820	5 684	6 649	5 861
澳大利亚 Australia	493	550	558	350	476	629
新西兰 New Zealand	178	235	242	128	187	192

资料来源：世界银行WDI数据库。

Data source: WDI database of World Bank.

11-17　集装箱吞吐量居世界前20位的港口
World Top Twenty Seaports in Terms to the Number of Containers Handled

单位：万标准箱　　　　(10 000 TEU)

港　口 Seaport	所属国家或地区 Country or Region	吞吐量 Containers Handled
上海 Shanghai	中国 China	3 174
新加坡 Singapore	新加坡 Singapore	2 994
香港 Hong Kong	中国 China	2 437
深圳 Shenzhen	中国 China	2 257
釜山 Pusan	韩国 Republic of Korea	1 618
宁波-舟山 Ningbo and Zhoushan	中国 China	1 472
广州 Guangzhou	中国 China	1 425
青岛 Qingdao	中国 China	1 302
迪拜 Dubayy	阿联酋 United Arab Em	1 300
鹿特丹 Rotterdam	荷兰 Netherlands	1 190
天津 Tianjin	中国 China	1 159
高雄 Gaoxiong	中国台湾 Taiwan, China	964
汉堡 Hamburg	德国 Germany	902
巴生 Kelang	马来西亚 Malaysia	890
安特卫普 Antwerp	比利时 Belgium	864
洛杉矶 Los Angeles	美国 United States	794
丹戎帕拉帕斯 Tanjung Periuk	马来西亚 Malaysia	750
厦门 Xiamen	中国 China	647
大连 Dalian	中国 China	640
长滩 Long Beach	美国 United States	609

资料来源：上海航运交易所，CL-ONLINE。
Source: Shanghai Shipping Exchange, CL-ONLINE.

11-18 港口货物吞吐量居世界前20位的港口（2010年）
World Top Twenty Seaports in Terms of the Cargo Handled, 2010

单位：百万吨 (million tons)

港 口 Seaport	所属国家或地区 Country or Region	吞吐量 Cargo Handled	备 注 Remark
上海 Shanghai	中国 China	653.4	内外贸货物 domestic and foreign trade cargo
新加坡 Singapore	新加坡 Singapore	501.6	内外贸货物 domestic and foreign trade cargo
鹿特丹 Rotterdam	荷兰 Netherlands	429.9	外贸货物 foreign trade cargo
天津 Tianjin	中国 China	413.3	内外贸货物 domestic and foreign trade cargo
宁波 Ningbo	中国 China	412.2	内外贸货物 domestic and foreign trade cargo
广州 Guangzhou	中国 China	411.0	内外贸货物 domestic and foreign trade cargo
青岛 Qingdao	中国 China	350.1	内外贸货物 domestic and foreign trade cargo
大连 Dalian	中国 China	314.0	内外贸货物 domestic and foreign trade cargo
香港 Hong Kong	中国 China	267.8	内外贸货物 domestic and foreign trade cargo
釜山 Pusan	韩国 Republic of Korea	263.0	内外贸货物 domestic and foreign trade cargo
秦皇岛 Qinhuangdao	中国 China	263.0	内外贸货物 domestic and foreign trade cargo
休斯敦 Houston	美国 United States	262.8	外贸货物 foreign trade cargo
南路易斯安娜 Southern Louisiana	美国 United States	223.3	内外贸货物 domestic and foreign trade cargo
深圳 Shenzhen	中国 China	221.0	内外贸货物 domestic and foreign trade cargo
黑德兰港 Headland Harbour	澳大利亚 Australia	199.0	内外贸货物 domestic and foreign trade cargo
蔚山 Ulsan	韩国 Republic of Korea	186.1	内外贸货物 domestic and foreign trade cargo
名古屋 Nagoya	日本 Japan	185.7	内外贸货物 domestic and foreign trade cargo
安特卫普 Antwerp	比利时 Belgium	178.2	内外贸货物 domestic and foreign trade cargo
巴生 Kelang	马来西亚 Malaysia	168.6	内外贸货物 domestic and foreign trade cargo
丹皮尔 Dampier	澳大利亚 Australia	165.0	内外贸货物 domestic and foreign trade cargo

资料来源：Shipping Statistics and Market Review, December 2011, ISL.

Source: Shipping Statistics and Market Review, December 2011, ISL.

11-19 海上商船拥有量居世界前20位的国家或地区
World Top Twenty Countries or Regions in Terms of the Number of Maritime Merchant Ships owned

国家和地区 Country and Region	艘数 （艘） Number of Vessels (unit)	载重吨 Deadweight ton	
		万吨 10 000 tons	占世界% Percentage in the World
世界总计 World Total	**47 833**	**134 893.0**	**100.0**
巴拿马 Panama	6 713	30 379.9	22.5
利比里亚 Liberia	2 604	16 270.6	12.1
马绍尔群岛 Marshall Islamds	1 443	9 583.5	7.1
中国香港 Hong Kong, China	1 633	9 212.2	6.8
希腊 Greece	1 106	7 126.8	5.3
新加坡 Singapore	1 585	6 588.9	4.9
巴哈马 Bahamas	1 202	6 178.9	4.6
马耳他 Malta	1 612	6 056.6	4.5
中国 China	2 663	5 135.4	3.8
英国 United Kingdom	902	3 567.8	2.6
塞浦路斯 Cyprus	843	3 220.0	2.4
日本 Japan	2 439	2 079.7	1.5
韩国 Republic of Korea	1 115	2 046.3	1.5
意大利 Italy	826	1 950.7	1.4
挪威 Norway	873	1 863.3	1.4
德国 Germany	477	1 787.5	1.3
印度 India	446	1 436.5	1.1
丹麦 Denmark	443	1 414.0	1.0
安提瓜和巴布亚 Antigua and Papua	1 245	1 388.2	1.0
印度尼西亚 Indonesia	2 314	1 158.7	0.9

注：1. 表中数据按载重吨排序；
2. 统计范围为300总吨及以上船舶，截至日期均为2011年1月1日；
3. 因统计口径不同，表中数据与其他出版物公布的数据略有差别。

资料来源：Shipping statistics and Market Review, Jannuang/February 2011. ISL.

Note: 1. The data in the table are arranged in the order of deadweight tons.
2. The ships, ranging over 300 tons and more in gross ton, are counted as of Jan. 1, 2011.
3. Owing to different statistical requirements, the data given in the table may be somewhat different from those issued in other publications.

Source: Shipping statistics and Market Review, Jannuang/February 2011. ISL.

11-20 集装箱船拥有量居世界前20位的国家或地区
World Top Twenty Countries or Regions in Terms of the number of Container Ships Owned

国家和地区 Country and Region	艘数（艘） Number of Vessels (unit)	集装箱位 Container	
		万标准箱 10 000 TEU	占世界% Percentage in the World Total
世界总计 World Total	**4 882**	**1 407.1**	**100.0**
利比里亚 Liberia	895	301.8	21.4
巴拿马 Panama	749	277.6	19.7
德国 Germany	297	123.1	8.7
中国香港 Hong Kong, China	276	97.0	6.9
新加坡 Singapore	328	86.1	6.1
英国 United Kingdom	198	79.2	5.6
安提瓜和巴布达 Antigua and Papua	404	57.4	4.1
马绍尔群岛 Marshall Islamds	212	52.7	3.7
丹麦 Denmark	94	51.3	3.6
中国 China	214	44.6	3.2
塞浦路斯 Cyprus	197	38.4	2.7
马耳他 Malta	95	28.5	2.0
美国 United States	85	26.8	1.9
希腊 Greece	32	18.7	1.3
法国 France	26	16.7	1.2
巴哈马 Bahamas	56	14.5	1.0
荷兰 Netherlands	69	10.5	0.7
意大利 Italy	21	8.0	0.6
韩国 Republic of Korea	70	6.4	0.5
印度尼西亚 Indonesia	117	5.7	0.4

注：1. 表中数据按集装箱位排序；

2. 统计范围为300总吨及以上船舶，截至日期均为2011年1月1日；

3. 因统计口径不同，表中数据与其他出版物公布的数据稍有差别。

资料来源：Shipping statistics and Market Review, Jannuang/February 2011. ISL.

Note: 1. The data in the table are arranged in the order of deadweight tons.

2. The ships, ranging over 300 tons and more in gross ton, are counted as of Jan. 1, 2011.

3. Owing to different statistical requirements, the data given in the table may be somewhat different from those issued in other publications.

Source: Shipping statistics and Market Review, Jannuang/February 2011. ISL.